Excel

YEAR 8

Science Revision Workbook

ESSENTIAL skills

Get the Results You Want!

PASCAL PRESS

Donna Bennett

Reprinted 2001, 2002, 2004, 2006, 2007, 2008, 2010, 2012

Updated in 2015 for the Australian Curriculum

Reprinted 2019, 2020, 2023, 2025

ISBN 978 1 74020 082 0

Pascal Press
PO Box 250
Glebe NSW 2037
www.pascalpress.com.au

Publisher: Vivienne Joannou
Project editor: Mark Dixon
Edited by Shelley Barons
Australian Curriculum updates edited by Geoffrey Thickett
Typeset by Scriptorium Desktop Publishing Pty Ltd and Kim Webber
Cover and page design by Dizign Pty Ltd
Printed by Vivar Printing/Green Giant Press

Contents

Introduction

This book is suitable for students studying the Year 8 Australian Curriculum Science course. It is particularly valuable for those students who are interested and motivated to seek an excellent result in Science.

This is a user-friendly revision workbook that students may use throughout the year to support classwork and help them with their revision for tests and exams.

The features of this book are as follows:

- The eight topics conform to the Australian Curriculum Science syllabus.
- Each topic consists of a concise summary of the key concepts in the topic.
- Each topic summary is followed by several pages of skill and knowledge practice questions to further enhance understanding.
- There are two sample Test Papers which bridge four topics and are aimed at the consolidation of knowledge.
- Answers to all questions are provided together with a complete glossary.
- Questions provide valuable revision before a test or exam.
- The book includes questions to interest the curious, capable student.

CHAPTER 1

LIVING THINGS

What would you get if you crossed an elephant with a kangaroo?

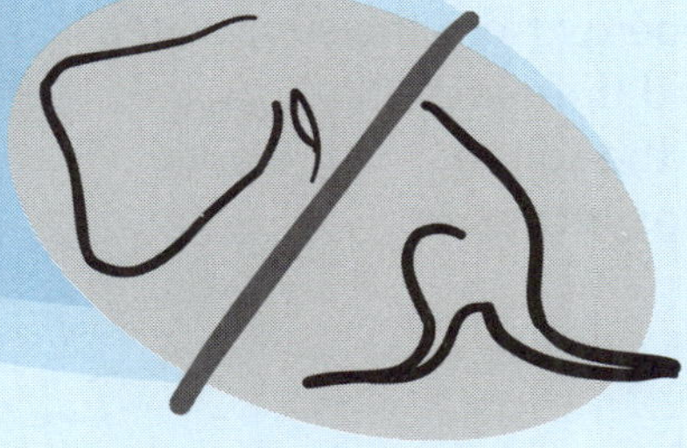

It's a bit difficult to define 'alive'. It is usually summed up by all the characteristics that living things share:

- **respiration** — the release of energy by chemical reaction.
- **taking in food** — Nutrition gives living things the energy and materials they need to move and make new cells (to grow, reproduce and repair).
- **excretion** — Waste body chemicals are poisonous and the body must get rid of them.
- **growth and development** — Living things get bigger and more complex with time.
- **sensitivity** — response to the world around them.
- **movement** — Animals move to find food. Plants don't need to do this but they can move some parts.
- **reproduction** — Living things eventually die so they must produce young.
- **photosynthesis** — Plants use light to make glucose from carbon dioxide.

All these things are called **life processes**. Animals and plants have all the life processes.

Is it alive?

To decide if something is **alive** we look at what it does. If we can find all the life processes, then it is alive, otherwise it is **non-living**.

For example, sparrows nest in the gutters of houses. The gutters are non-living but the sparrows are alive.

Working together

Organisms, both plant and animal, are made up of billions of cells all with their own job to do.

- Cells that do the same job are often grouped together into a **tissue**, such as muscle or nerve tissue. Different tissues join together to make **organs**, such as lungs. Different organs do different jobs.
- If several organs work together to do a job its called a **system** (eg reproductive system).
- Specialised cells do a specific job. The cells in your nose have little hairs called cillia that move particles to the back of the throat where they are swallowed then destroyed by the stomach acid. Root cells have tiny hairs too. These increase the surface area which helps the root take in water and minerals.

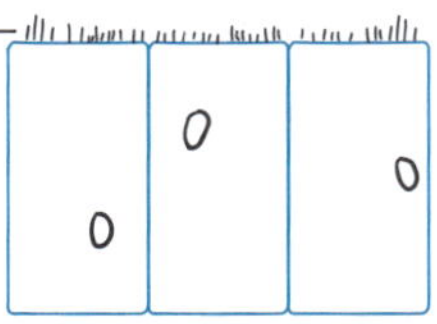

Cilliated tissue

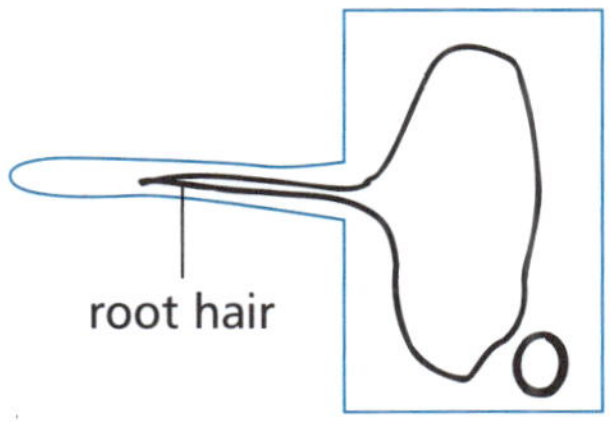

Root hair

Important parts of plants

Plants are different from animals because they can make their own food. They use the energy from the sun, water, air and nutrients from the soil to **photosynthesise**.

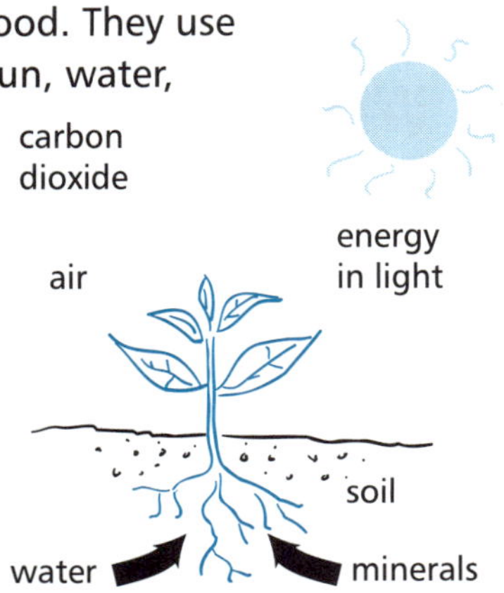

What a plant needs to make food

Plants have many organs:

- **Leaves** — take in the gases (carbon dioxide) the plant needs, and get rid of waste and surplus H_2O.
- **Stem** — gives the plant structure and supports the leaves, flowers etc.
- **Roots** — take in water and minerals and hold the plant in the soil.
- **Buds** — grow into new flowers or leaves.
- **Flowers** — used to reproduce the plant.

Plants have tissues too. **Xylem** is a plant tissue. Xylem cells join together to form long tubes inside the plant. These tubes transport water to all parts of the plant. **Phloem** is a tissue that carries food from the leaves to the rest of the plant.

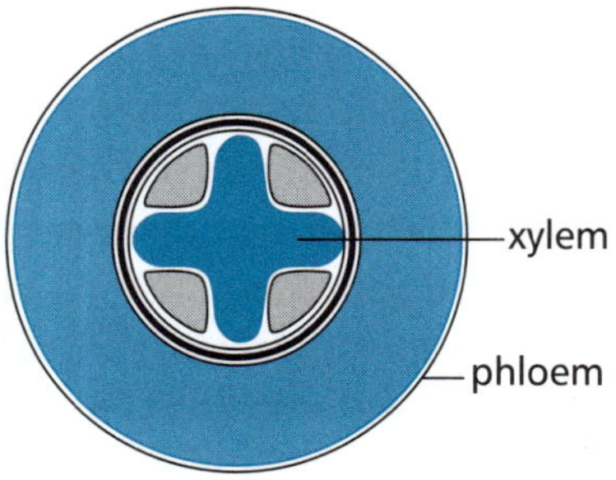

Section through a root

Important parts of animals

- **Nervous system** — This system enables animals to react to their surrounding environment. Nerves send messages from one part of the body to another.
- **Sensory system**

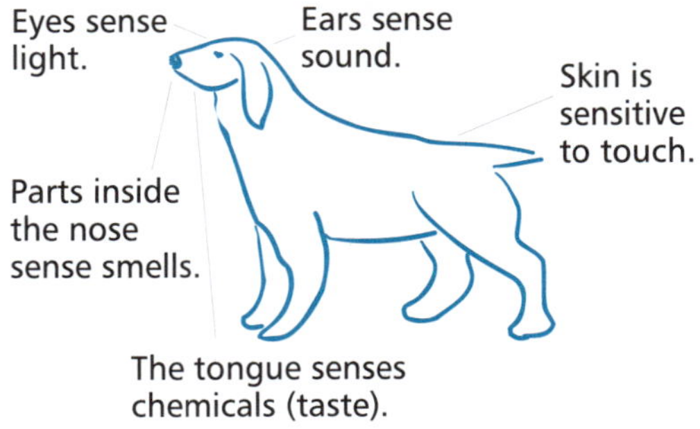

- **Digestive system** — The digestive system breaks down food into simple substances that can be used by the body.
- **Circulatory system** — This system carries blood around the body. Blood transports nutrients and wastes.
- **Respiratory system** — This is a gas-exchange system. Animal cells need to take in oxygen and give out carbon dioxide.

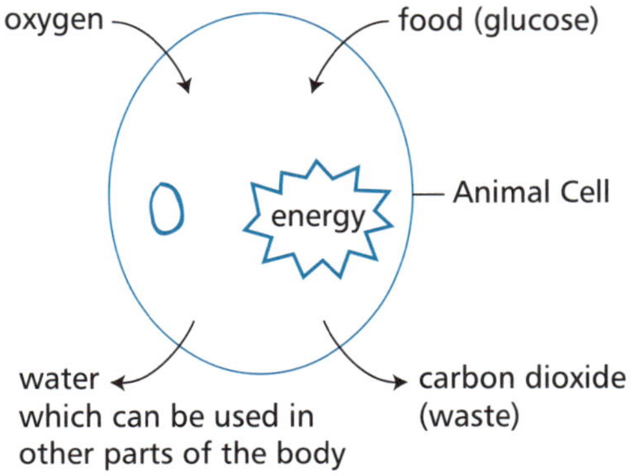

Respiration happens in cells

- **Skeletal system** — A framework of bones that supports an animal's body.
- **Muscular system** — Tissues that allow animals to move around.
- **Reproductive system** — All living things have the capacity to increase their numbers — produce offspring.
- **Excretory system** — Humans excrete sweat, carbon dioxide and urine. Faeces is also excreted, but is **undigested** waste. It passes straight through the body through a long tube from the mouth to the anus, without ever being absorbed and used by the body.

Starting life

All living things grow and change during their lives.

Many animals and plants start their lives as two specialised cells (sex cells or gametes).

For a new plant or animal to grow a male and female sex cell must join together. This is called **fertilisation**.

For animals

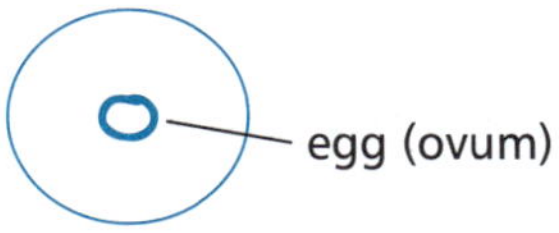

Female sex cell (gamete)

Male sex cell (gamete)

For flowering plants

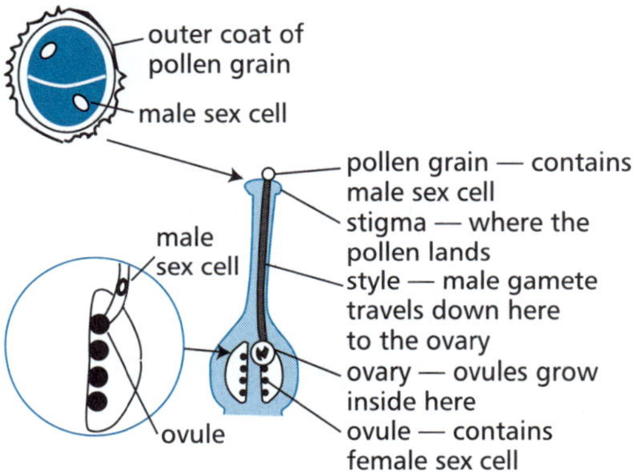

Female sex cell (inside the ovary)

Male sex cell (inside the pollen grain)

Flowers have many parts and each part has a job to do.

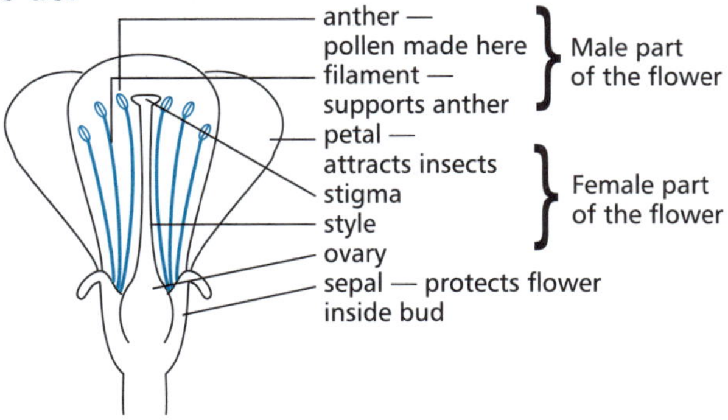

How living things develop depends partly on what they inherit from their parents and partly on their environment.

Humans breed plants and animals with the characteristics we desire. For example, we choose to grow the seeds from the biggest, juiciest watermelon. This is called selective breeding.

Using biological keys to identify living things

A key is a listing of specific characteristics such as behaviour and structure by which a living thing can be identified.

A key operates by selecting features which differ for each organism.

Using simple keys

Go to the start of the key below. When you come to a branch choose which way to go.

To use the key to identify the animal, the bacteria, the insect and the plant.

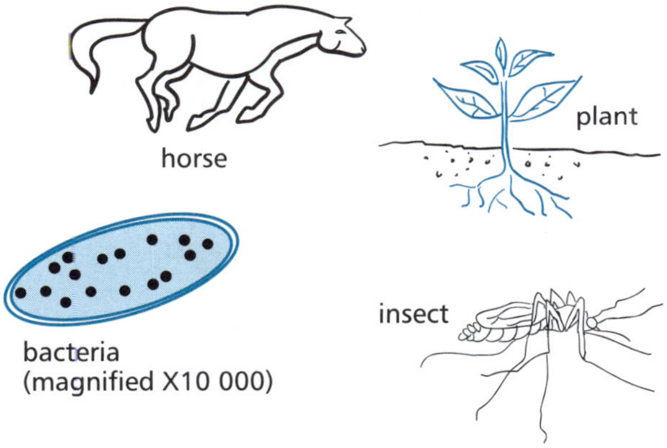

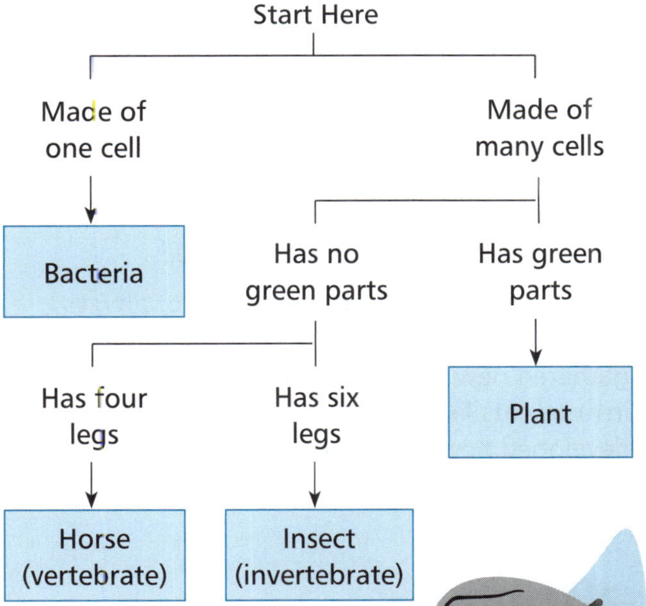

A key can also be written as a list of questions.

1 Made of one cell . . . . **Bacteria**

Made of many cells . . . . go to 2

2 Has green parts . . . . **Plant**

Has no green parts . . . . go to 3

3 Has four legs . . . . **Horse**

Has six legs . . . . **Insect**

All the main groups can be divided into smaller groups using a key.

There are far too many living things for one key so we use separate, smaller keys for different groups. For example, we could use a key just for winged insects or for flowers with five petals.

The following is a key for this group of bugs.

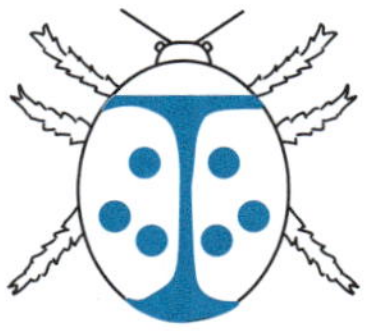

bug A

bug B

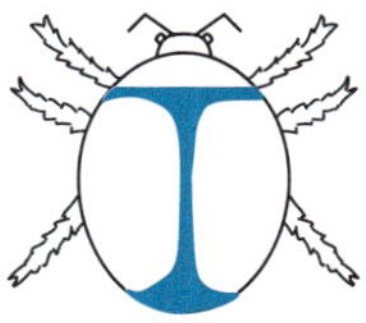

bug C

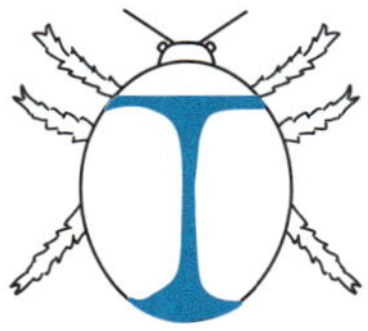

bug D

1 Spotted bug . . . . go to 2

Non-spotted bug. . . . . go to 3

2 Many spots . . . . Bug B

Few spots . . . . Bug A

3 Straight antennae . . . . Bug D

Bent antennae. . . . . Bug C

Remember that living things have not always been classified in the same way. Scientists have gathered new information using scientific instruments like the microscope and have developed new ideas like the theory of evolution.

Glossary

Digestion
The breakdown of large, insoluble food molecules into small, soluble ones that can be absorbed.

Fertilisation
The process whereby male and female sex cells join.

Gametes
The male and female sex cells.

Life processes
These are all the things that living things can do — move, sense, grow etc.

Organs
Groups of tissues joined together and working together.

Photosynthesis
The process in which plants use light energy to make glucose from carbon dioxide.

Reproduction
When living things produce young of the same kind as themselves.

Root hairs
The fine extensions of plant roots through which water and minerals are absorbed.

Tissue
A group of similar cells grouped together.

Science skills

Question 1

a Complete the table below with the words **yes** or **no**.

Does it ...?	Plant	Soil
Reproduce		
Get rid of waste		
Take in carbon dioxide		
Produce food		
Grow		
Move		
Sense things		

b Is the soil alive? Is the plant alive?

Question 2

Different parts of plants do different jobs.

a Complete the table below.

Part of a plant	What is its job?
flower	
stem	
leaf	
root	
bud	

b Plants make their own food. What do they need to do this?

Question 3

Label the parts of the digestive system with the following words: oesophagus, stomach, small intestine, large intestine, rectum, anus.

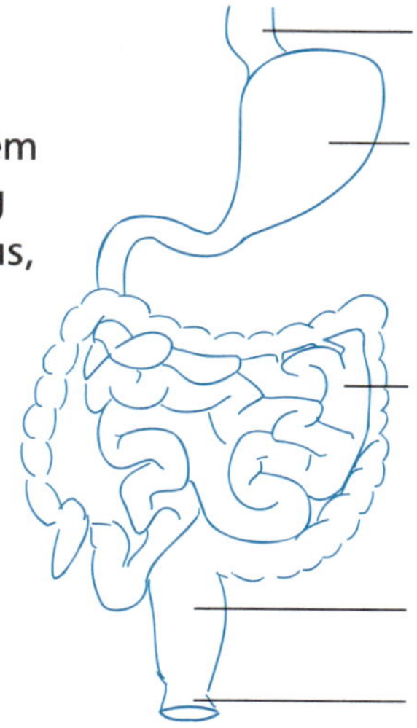

Question 4

The nervous system receives messages from the senses and transfers them to the brain.

Fill in the names of the five senses.

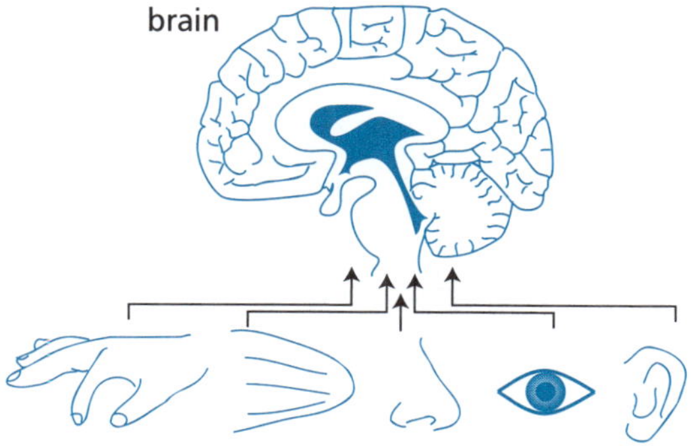

Question 5

Living things must have food for energy and the materials they need to move and make new cells. Where do they get their food from?

Living things	How they get their food
Bacteria	
Plants	
Animals	

Question 6

What jobs do flower parts do? Fill in the function (what it does) of the flower part and tick 'Male' or 'Female'.

Part	Male	Female	Function
anther			
filament			
stigma			
style			
ovary			
petal			

Question 7

Use the information below to complete this key.

Vertebrate Groups

fish ________ ________ bird ________

eg ________ ________ brown snake ________ ________

Sparrows have feathers and lay eggs.

Toads have moist skin with no scales.

Brown snakes have a dry, scaly skin and lay eggs with tough shells.

The **blue whale** is a mammal. It feeds its young on milk.

Fish have scales, fins and gills.

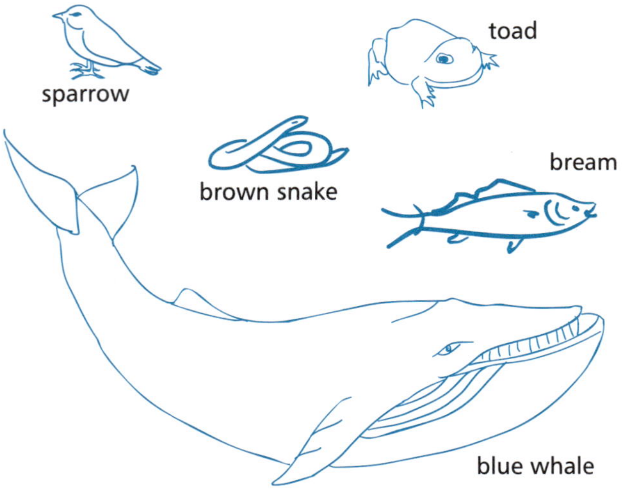

Question 8

Flowering plants produce seeds but more primitive plants, like ferns and moss, produce spores.

Look at the plants below and draw a simple key to work out which plant group they belong to.

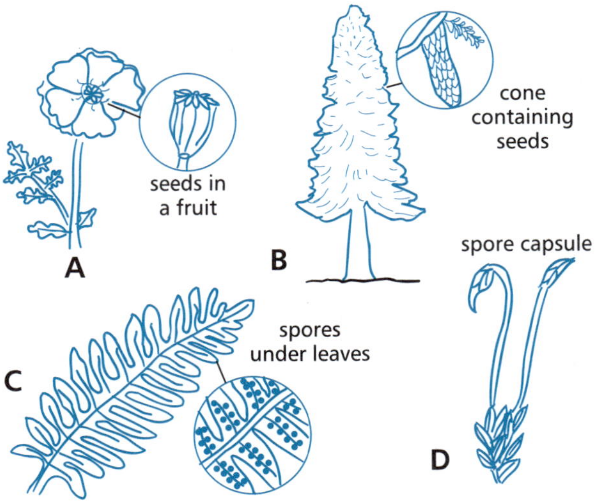

Question 9

The shells of molluscs are some of the most fascinating shapes produced by organisms.

Study the information below and match each sample shape with one of the three groups.

Whelk — With every turn, the inside diameter of the shell gets bigger, producing a shell with a short spiral.

Topshell — The shell has an inside cavity, and its sides are flattened, not rounded.

Cerith — With each turn, the inside diameter of the shell increases only a little. So we have a long, thin, tightly-wound shell.

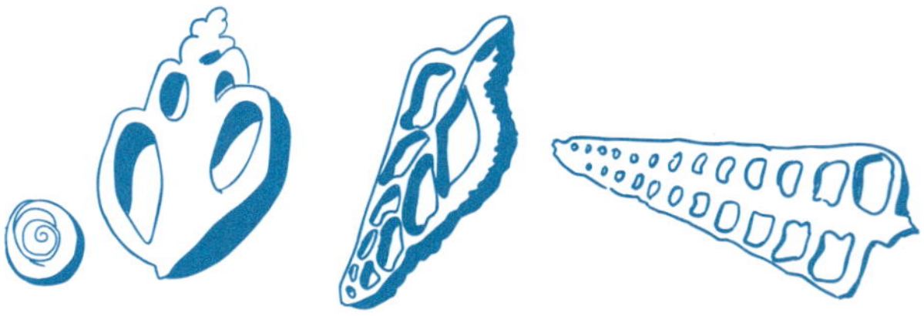

Sample Mollusc Shapes

Question 10

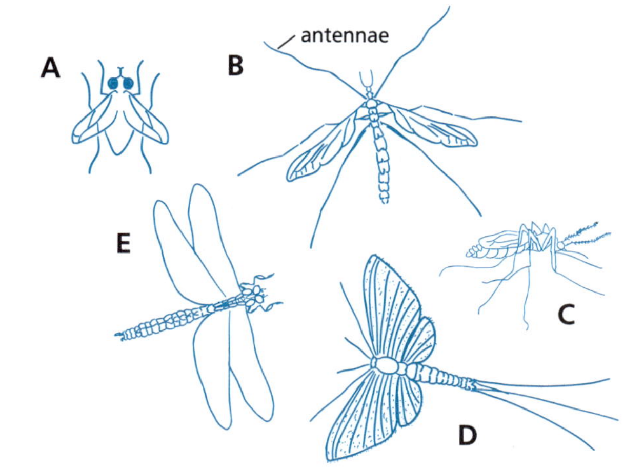

Use the simple key below to identify these flying insects.

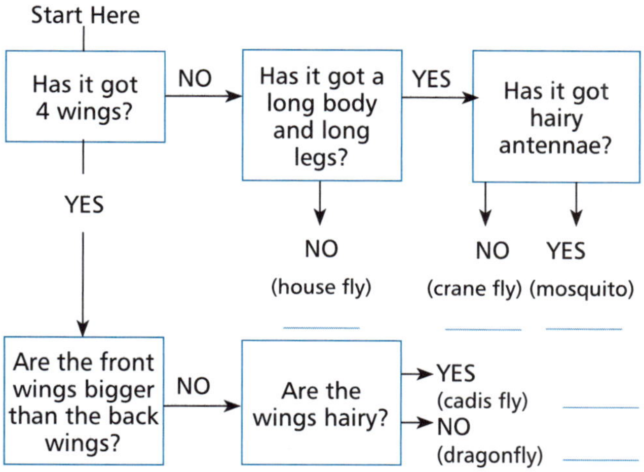

Question 11

Prepare a question key for the group of fish below. The first one is done for you.

1	Fish with a long tube like body	**Fish 1**
	Fish with a regular body shape	go to 2

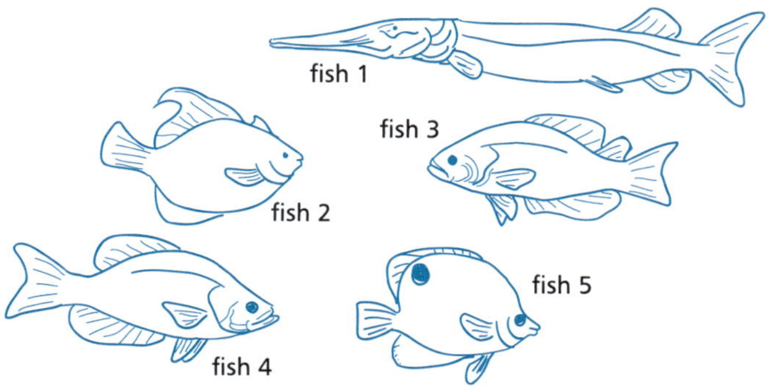

Question 12

Prepare a simple branched key for these needles from several pine trees.

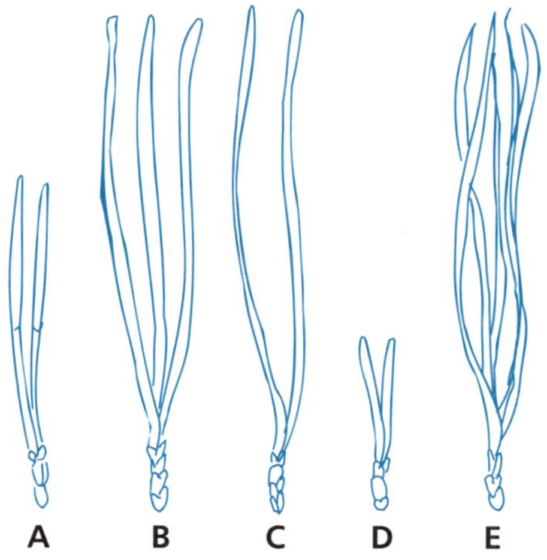

Question 13

Use the following key to find out what sort of flowers these are.

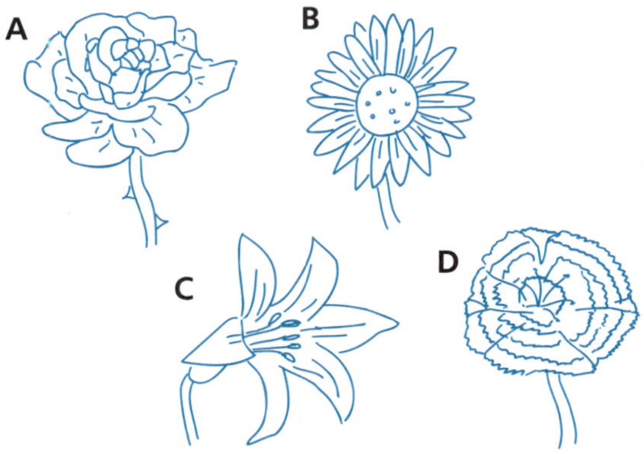

1	Has many petals	go to 2
	Has only a few petals	Lily
2	Has pointed petals	Daisy
	Petals not pointed	go to 3
3	Stem has thorns	Rose
	Stem does not have thorns	Carnation

Question 14

Put the sentences in the correct order to complete the flow chart.

The male sex cell moves down the pollen tube into the ovule.

Fertilisation occurs.

The pollen tube grows down to an ovule.

A pollen tube grows down through the style.

Flow Chart

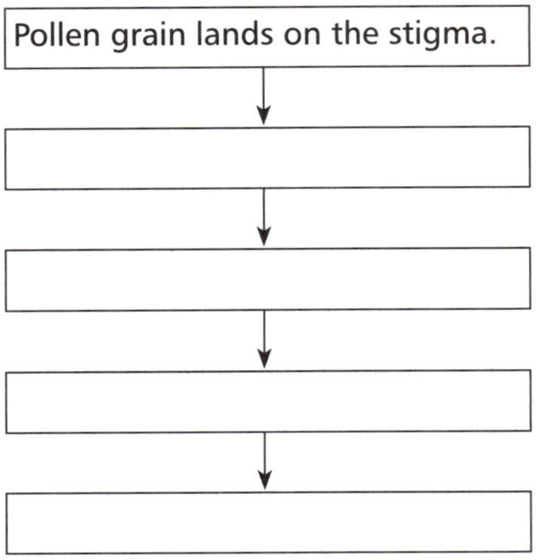

Question 15

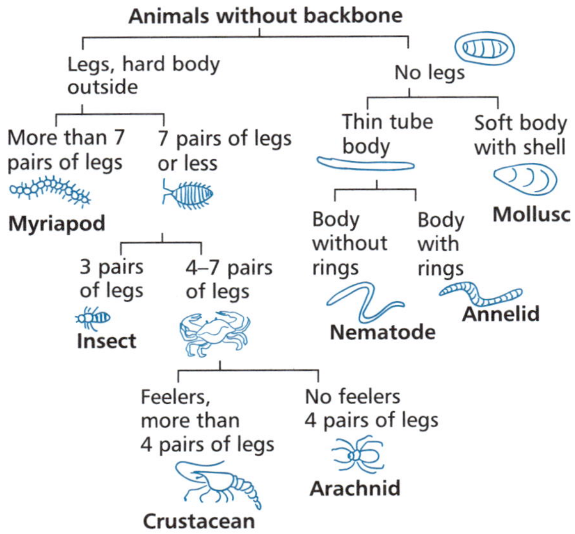

a Use the key to write a list of the characteristics of arachnids.

____

____

b Which characteristics of arachnids are also the characteristics of crustaceans?

____

____

c What makes arachnids different from crustaceans?

____

____

d What do nematodes and annelids have in common?

____

____

e How could you distinguish between a nematode and an annelid?

____

____

Revision questions

Question 1

Plants cannot move about like animals but parts of plants can move. Carnivorous plants, such as the Venus flytrap, trap insects by moving certain parts.

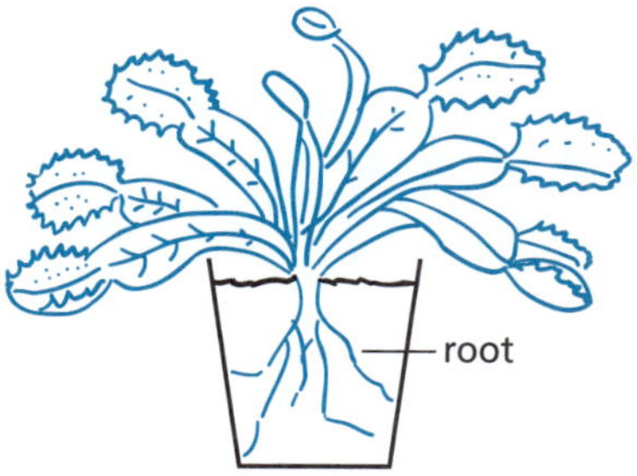

a Which part of the plant moves?

____

b Why does this happen?

____

____

c Why can't the Venus flytrap move about?

____

____

Question 2

Complete the sentences with the correct words.

Plants and ____________ are living organisms. They ____________, ____________, ____________, ____________, and get rid of ____________. Plants make their own food using ____________ from sunlight, ____________ and ____________ from the soil and ____________ ____________ from the air. Different parts of a plant do different ____________. The different parts are called ____________.

Question 3

a Different parts of your body are called ____________. These work together in groups called ____________.

b Complete the table.

System	Function
	Gives you information about the environment
	Breaks down food
	Carries food, oxygen and water around the body

Question 4

Look at the animals in the pictures below.

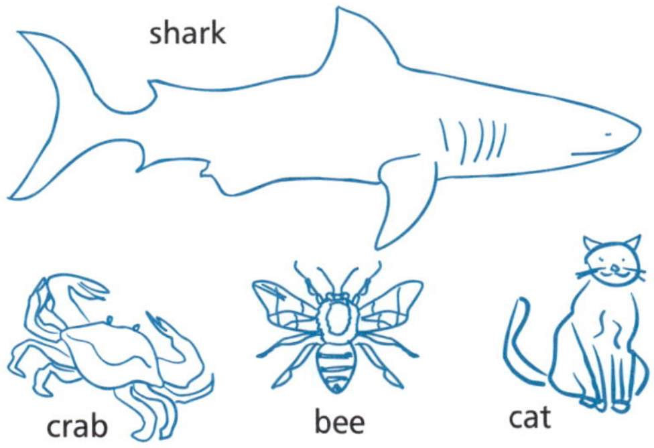

a List them in order of how many cells they have.

____

b What do we call an organism which is only one cell?

____

Question 5

Complete the sentences.

____________ tubes carry ____________ to all parts of a plant and ____________ tissue carries ____________ from leaves to all other parts of the plant.

Question 6

a What five things can a cat sense?

____

b What two things do plants sense?

c Where in plants and animals does respiration occur?

d What waste product is produced when animal and plant cells respire?

e What waste product is produced when plants photosynthesise?

f What three things do you excrete from your body?

Question 7

Can you think of a non-living thing that moves around, uses oxygen to release energy, and flies?

Question 8

Label the diagram of the plant below using the words supplied.

flower, bud,
leaf, stem,
root,
root hair

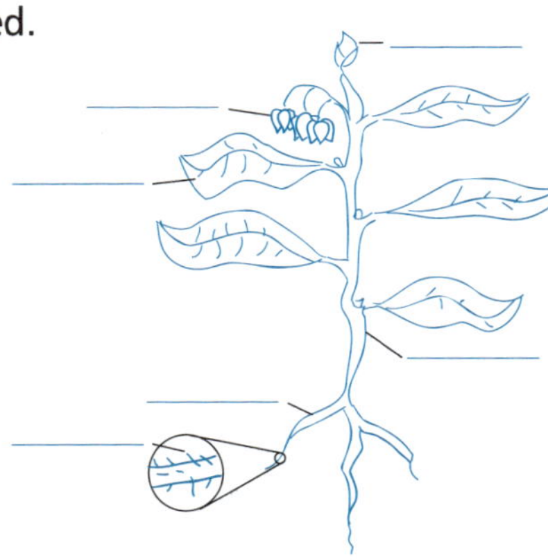

Question 9

Label the diagram of a flower using the word list below.

petal, anther,
stigma, sepal,
ovary, filament,
style

Question 10

a A terrestrial animal was found. It had no internal or external skeleton but it did have a segmented body. Into which phyla could it be placed? Circle the correct answer.

A Mollusca   C Arthropoda
B Annelida   D Echinodermata

b A biologist discovered a four-legged, seed-eating animal that sleeps under logs and rocks during the day and has a body temperature which always remains at 36°C. The animal appears to be:

A an arthropod   C a mammal
B a reptile   D none of these

Question 11

a Draw a simple key to separate the following spiders.

b Draw a question key to separate the spiders.

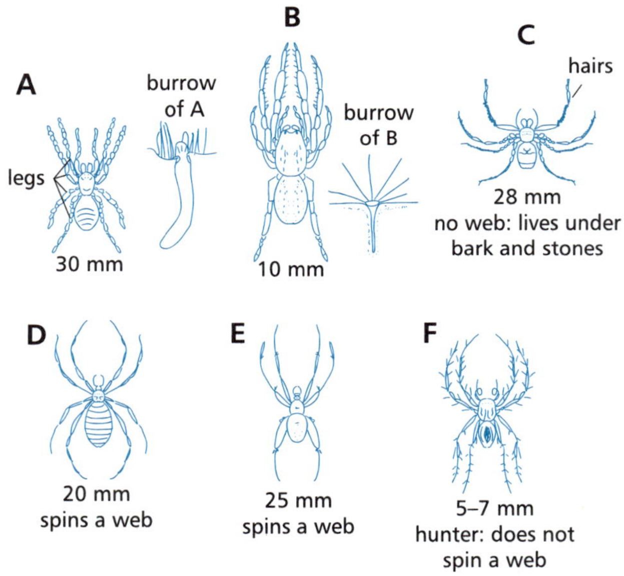

Question 12

Look at the drawings of two cats. Use the key to find out the cats names.

Key

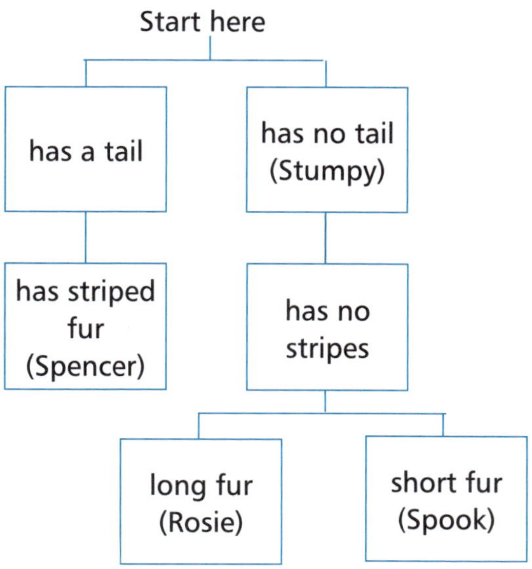

Question 13

Use the diagrams to complete the key below.

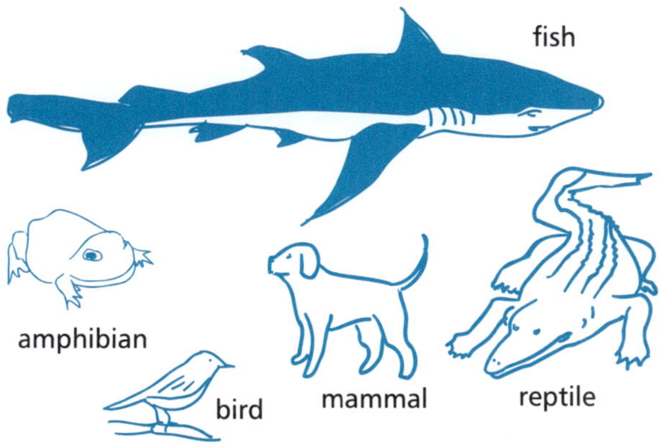

1. Animal has fins ________

   Animal does not have fins go to 2

2. Animal lays eggs go to 3

   Animal does not lay eggs ________

3. Animal has feathers ________

   Animal does not have feathers ________

4. ________________ ________

   ________________ ________

Question 14

a Study the pictures below and list the life processes that each thing has.

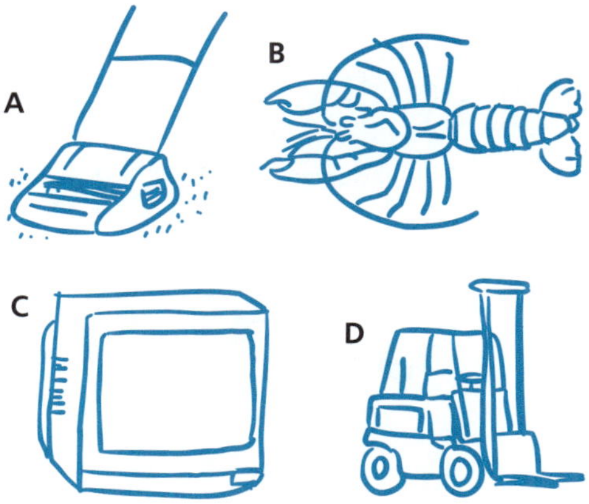

Life Processes — sense, move grow, reproduce, use oxygen, get rid of waste

____

____

____

____

b Which ones are living animals?

____

Question 15

Add the following words to the correct space in the sentences below:

oxygen, waste products, food.

The circulatory system carries ____________ and ____________ to all parts of the body. It also carries ____________ ____________ away from all parts of the body.

Question 16

Write down three ways that a dog, a cat and a rabbit are like each other but different from a snail.

____

____

____

Question 17

Use the following key to find out which group organisms A, B, C and D belong to.

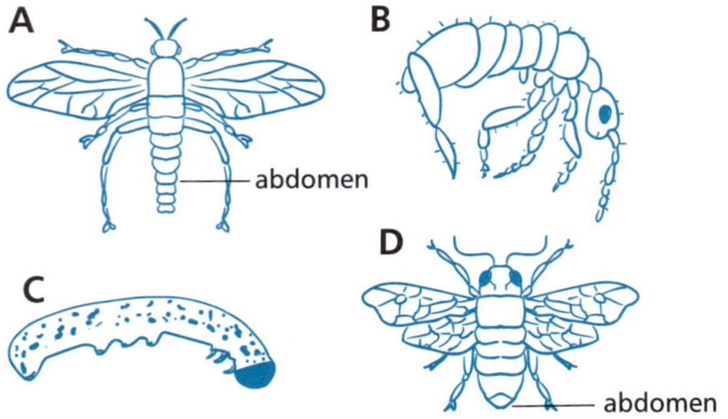

1. Wings present	go to 2
Wings not present	go to 3
2. Abdomen fewer than six segments	Soldier flies
Abdomen more than six segments	Sawflies
3. Three pairs of legs	Springtails
More than three pairs of legs	Grass grubs

____

____

____

____

Question 18

Look at the diagram of a biceps muscle below and then fill in the missing words.

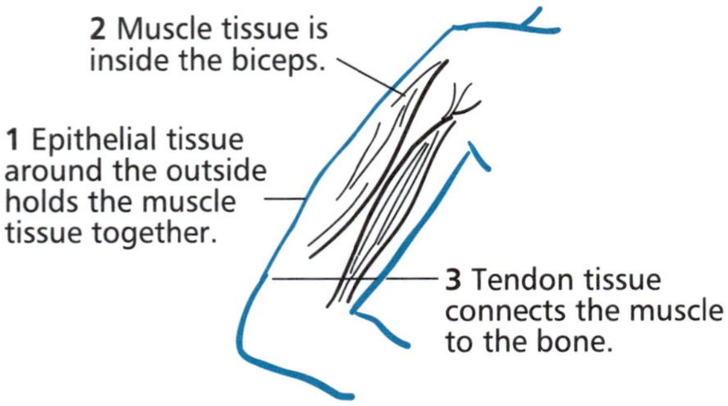

My bicep is an ____________, formed from several tissues joined together. Tissues in the biceps include ____________, ____________ and ____________ tissue.

Question 19

a In what ways is water gained by the body?

____

____

b During normal temperatures (20ºC), water is lost by the human body:

i most ____________

ii least ____________

Question 20

List two major criteria that are best used to group living organisms and then to draw a classification key.

____

____

CHAPTER 2

ANIMAL SYSTEMS

What happened to breakfast?

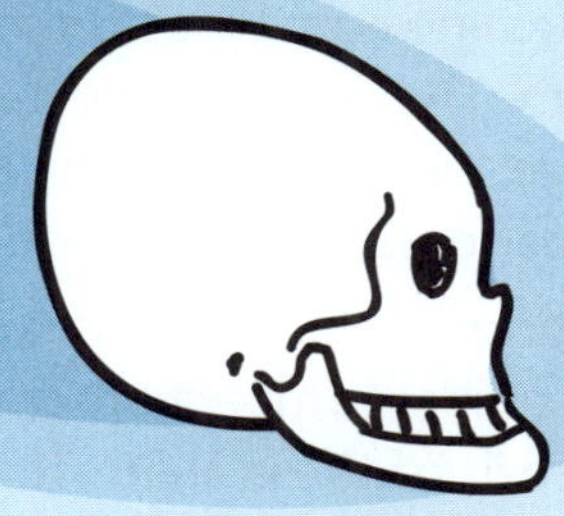

A system is an ordered collection of processes, with inputs and outputs, which has an identity separate from the rest of the animal. However, animals only survive if all their systems are co-ordinated and working together.

The human body has many different types of cells. Each of them is specialised to do a particular job. Some specialised cells are grouped together to make tissues. Tissues are grouped together to form organs and groups of organs make up systems.

Humans have many systems in their bodies — for example the skeletal system, circulatory system, digestive system and respiratory system.

Digestive system

The digestive system is a series of organs that break food down into simple substances like sugars, amino acids and lipids, that diffuse into cells for the body to use.

A tube called the alimentary canal, or gut, runs from the mouth to the anus. All vertebrates, including humans, have similar digestive systems.

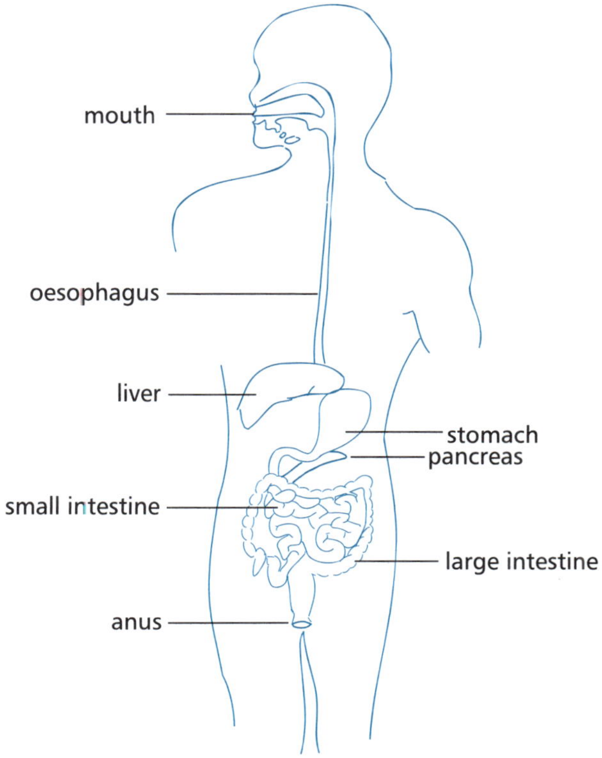

Human digestive system

Parts and functions of the digestive system

- **Mouth** — This is the entry to the digestive system. Food is cut up into smaller pieces by the teeth. **Saliva** contains the enzyme, **amylase**, which begins to break down carbohydrates into simple sugars.
- **Oesophagus** — This is a muscular tube that leads from the mouth to the stomach. When an animal swallows food, the muscles contract and move the food towards the stomach. This is called **peristalsis**. Food passes through the whole gut this way.
- **Stomach** — Powerful muscles churn up the food to break it into smaller bits. Hydrochloric acid starts to digest the food and the enzyme pepsin starts to break down the protein. A circular muscle at the base of the stomach called the pyloric sphincter keeps food in the stomach until it is ready to move on.

- **Small intestine** — This is a long tube in which the food is digested and absorbed. In humans the small intestine comprises the **duodenum** and the ileum. The lining of the duodenum produces digestive juices and the lining of the ileum absorbs food into the blood.
- **Large intestine** — This absorbs water from the residual solids, compacts them, and controls the exit of these solids through the anus.
- **Anus** — Exit point for unwanted food.

Organs that support the digestive system

- **Pancreas:**
  - secretes enzymes into the small intestine to break down protein, carbohydrates and fats
  - secretes bicarbonate which neutralises stomach acid
  - produces hormones which act on the liver to regulate blood glucose
- **Liver:**
  - supplies energy quickly
  - stores vitamins and minerals
  - regulates blood clotting
  - detoxifies poisons

Circulatory system

The circulatory system constantly carries blood in cycle around an animal's body. This system supplies every cell with oxygen and the nutrients it needs to survive, and takes away the waste products.

The blood flows through tubes called blood vessels and is kept moving by the pumping action of the heart.

Parts and functions of the circulatory system

- **Heart** — The heart is a hollow muscular pump which forces blood around the body. An **atrium** is a chamber which admits blood to the heart and a **ventricle** expels blood from the heart.

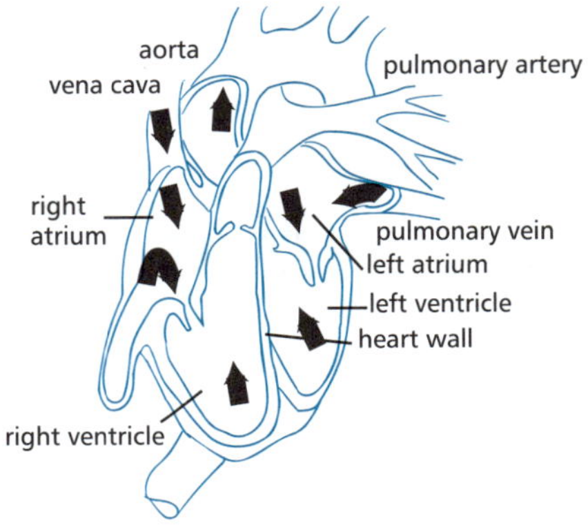

Human heart

Blood from the lungs flows into the heart via the **pulmonary vein**, and out to the body via the **aorta**. Blood from the body flows back into the heart via the **vena cava**, and then out to the lungs via the **pulmonary artery**.

- **Blood vessels** — An **artery** is a thick muscular tube that carries blood, under pressure, away from the heart. The walls of an artery stretch and contract creating a **pulse**. **Capillaries** are very fine tubes that carry oxygenated blood to every cell in the body. They have very thin walls allowing substances to pass from the blood to the cells. After blood has passed through the capillaries it goes to fine vessels called venules which join up to form thicker walled veins which carry the deoxygenated blood back to the heart.
- **Blood** — Blood is a fluid that carries oxygen and nutrients from digested food to the body's cells and carries away the waste. Blood is made up of:
  - **plasma** (the fluid part of blood)
  - **red blood cells** (contain haemoglobin which carries the oxygen),
  - **white blood cells** (combat infection)
  - **platelets** (cell fragments that help blood to clot).

Excretory system

The excretory system is responsible for eliminating waste products such as carbon dioxide, water, salts, and nitrogenous waste.

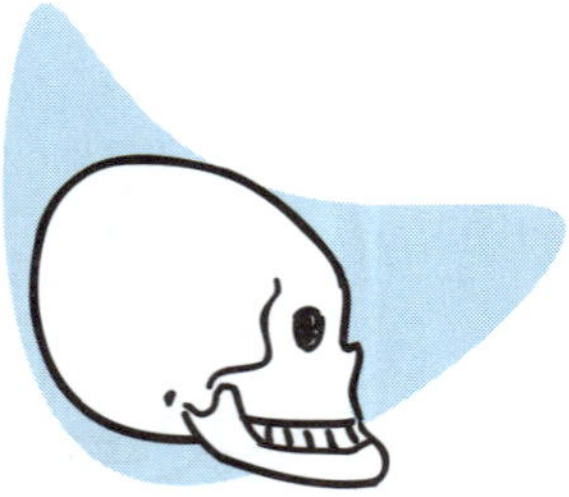

Parts and functions of the excretory system

The main excretory organs in vertebrates are the skin, lungs and kidneys.

- The skin plays an important role in maintaining body temperature through **sweating** and **shivering**.
- The lungs return unwanted carbon dioxide and other gases to the atmosphere.
- The kidney filters waste from the blood and controls body water content. The waste is concentrated into a product called **urine** which passes to the bladder via the ureter and is excreted.

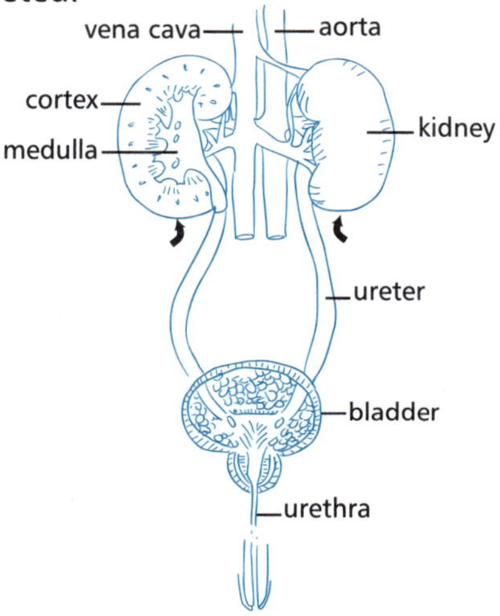

Human excretory system

Skeletal system

The skeleton is a framework that supports the animal's body and provides an anchor for its muscles. The skeleton is made of bone and allows animals to move.

Parts and functions of the skeletal system

- **Backbone** — A flexible chain of bones called **vertebrae** known as the spine. Together the vertebrae form a hollow rod which protects the **spinal cord**.
- **Ribcage** — A bony cage, made of curved bones called ribs, that allows breathing and protects internal organs.
- **Bones for movement** — The bones of the girdles (hip and shoulders) and upper and lower limbs (arms, legs, flippers, wings). The bones an animal uses for locomotion.
- **Hindlimb** — This has three main bones: femur (thighbone), tibia and fibula (lower hindlimb). These bones meet at the knee joint.
- **Forelimb** — This has three main bones: the humerus (upper arm), and the radius and ulna (lower arm). The upper and lower arm meet at the elbow.
- **Hand** and **foot** — The hand is used for gripping and the foot for movement. The hands have bones called carpals, metacarpals and phalanges. The foot bones are called tarsals, metatarsals and phalanges.

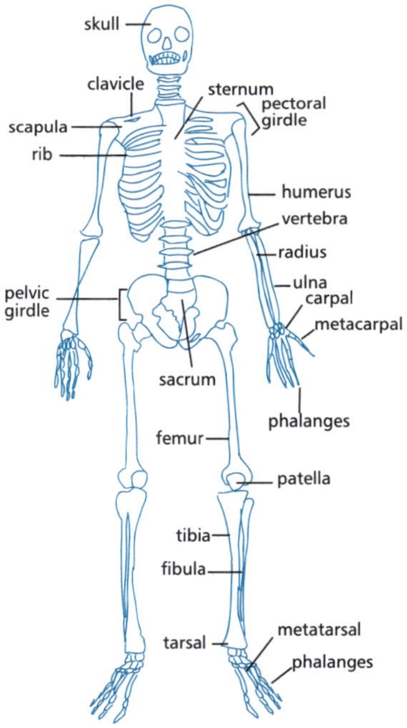

Human skeleton

- The place where two bones meet is called a joint. Some joints are fixed (unable to move), some are slightly moveable and some are freely moveable like hinge and ball and socket joints (synovial joints).

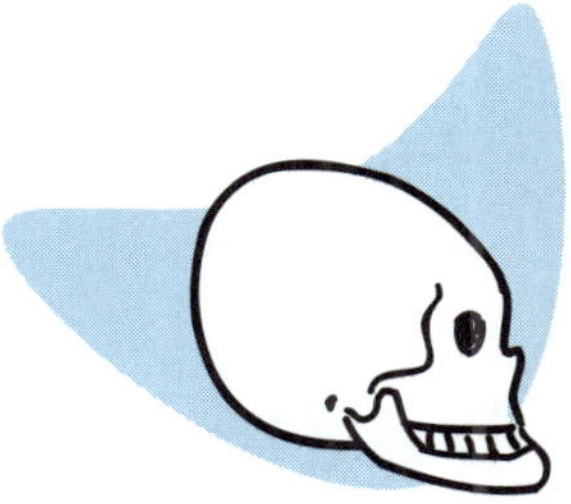

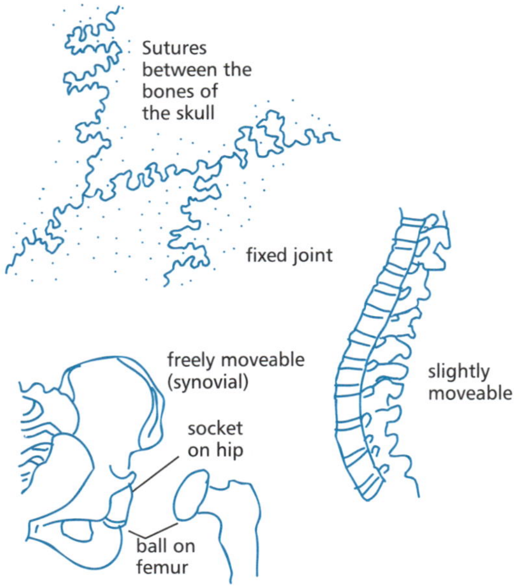

Respiratory system

The lungs are the organs at the centre of the respiratory system. They bring air and blood close together so that oxygen and carbon dioxide can move from one to the other through a thin, moist membrane. Breathing is the forced movement of air into and out of the lung. When the carbon dioxide levels in the brain decrease the diaphragm muscles pull down and so increase the volume of the lungs, so air rushes in. The diaphragm then contracts and the air is forced out.

Parts and function of the respiratory system

- **Lungs** — Lungs are the organs that most vertebrates use to exchange gases. Oxygen and carbon dioxide exchange occurs through a very thin membrane.

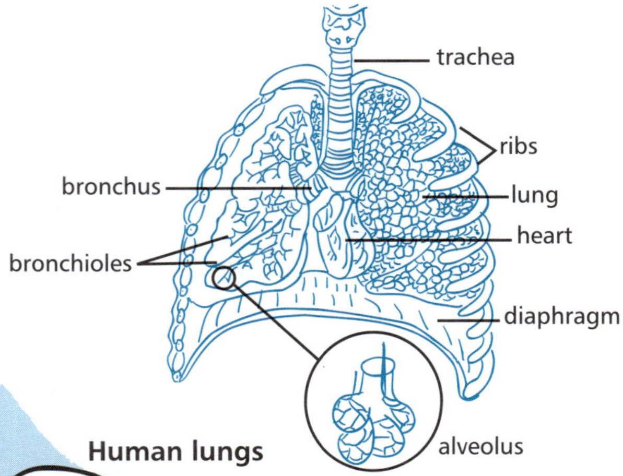

Human lungs

- **Trachea** — The passageway leading from the nose to the lungs. The trachea is reinforced by rings of cartilage and lined by cilia that move dust and other particles back out of the airways.
- **Bronchus** — The passageway leading from the trachea to the lung. The bronchi branch into very small branches called bronchioles which go to every alveolus.
- **Alveolus** — A tiny air-filled sac from which gas exchange between blood and air takes place across a thin membrane.

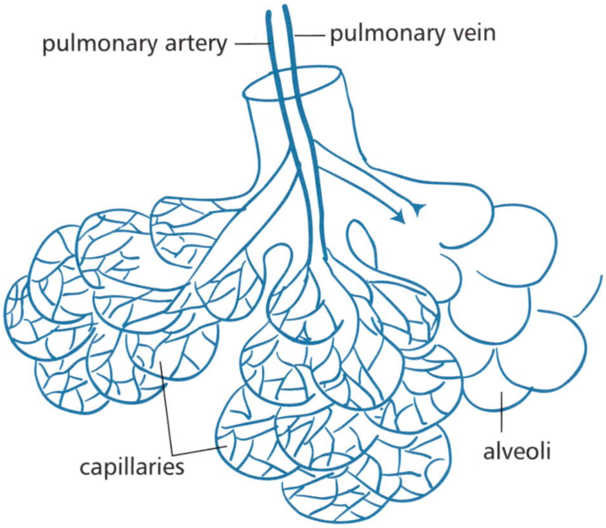

Glossary

Alveolus
A tiny sac in which gas exchange takes place.

Amylase
An enzyme in saliva that breaks down carbohydrates into their sugars.

Artery
A blood vessel that carries blood away from the heart.

Atrium
A chamber that admits blood to the heart.

Capillaries
Microscopic blood vessels that penetrate the tissues carrying sugars and oxygen to individual cells.

Duodenum
The first section of the small intestine where acid from the stomach mixes with digestive juices from the pancreas and liver.

Peristalsis
Rhythmic waves of contraction of smooth muscle that push food along the digestive tract.

Pulse
The momentary expansion of an artery following a heartbeat.

Saliva
A watery liquid produced by the salivary glands which lubricates food in the mouth, beginning carbohydrate digestion.

Shivering
The repeated contraction of muscles to produce heat.

Spinal cord
A column of nervous tissue that runs through the backbone.

Sweating
The secretion of moisture from the skin to cool the body.

Ventricle
A chamber that expels blood from the heart.

Vertebra
One of the bones that makes up the backbone.

Vein
A blood vessel that carries blood back to the heart.

Question 1

Label the diagram of the human skeleton below with the following words:

skull, sternum, humerus, radius & ulna, carpal, metacarpal, phalanx, femur, patella, tibia & fibula, tarsal, metatarsal, pectoral girdle, clavicle, scapula, rib, sacrum, vertebra, pelvic girdle

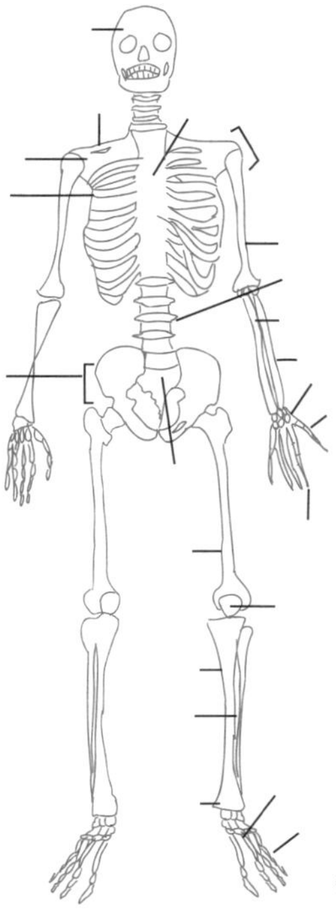

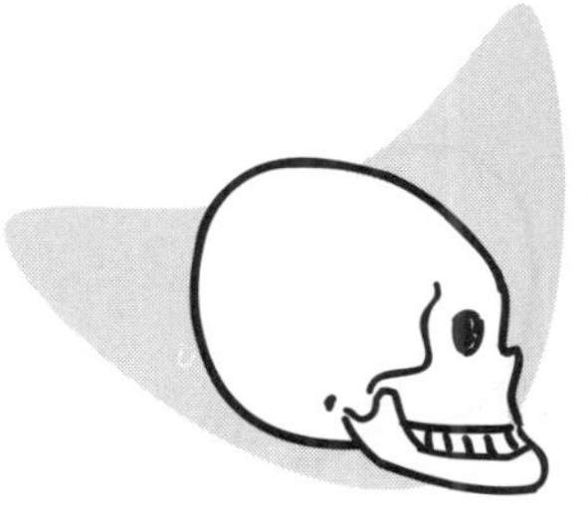

Question 2

The following diagram is the forelimb of a horse. Label the following bones:

scapula, humerus, radius, carpus, metacarpus, phalange (digit)

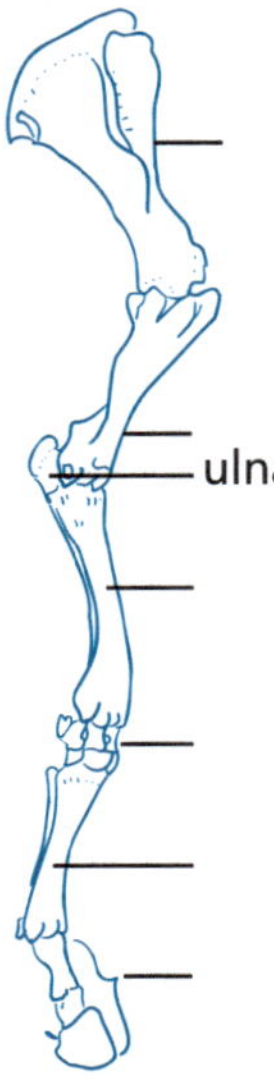

Question 3

Label the hindlimb of a rat with the following terms:

pelvic girdle, femur, patella, fibula, tibia, tarsus, metatarsus, phalanges

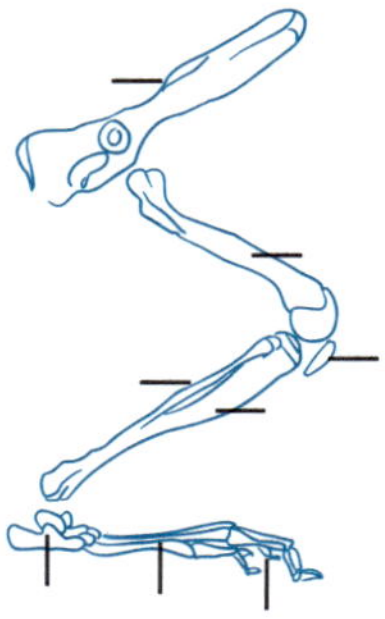

Question 4

Complete the following table which lists the scientific names and common names of some human bones.

Scientific name	Common name
clavicle	
mandible	
	breastbone
	hip bone
metatarsals	
	shin bone
patella	
	shoulder blade

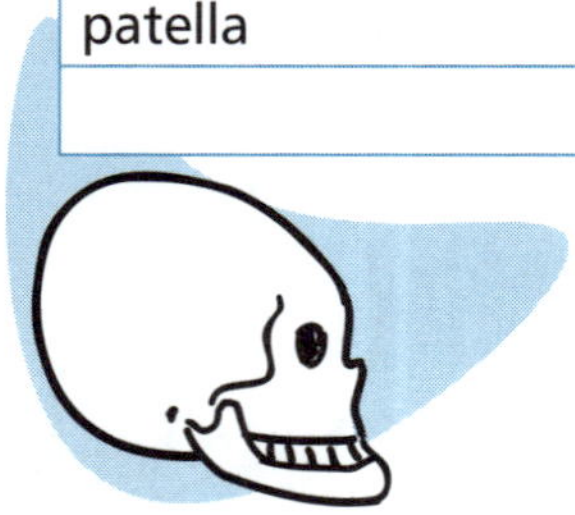

Question 5

Label the diagram below using the following words:

artery, heart, vein

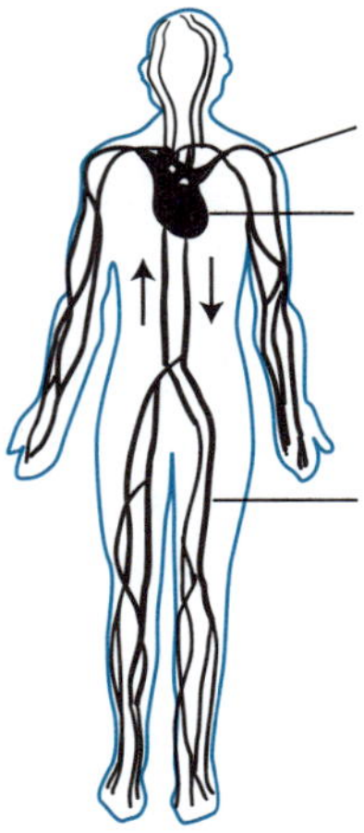

Question 6

Study the diagram of the heart and complete the table matching the labels below with the correct letters in the diagram. Write the function of each part in the third column.

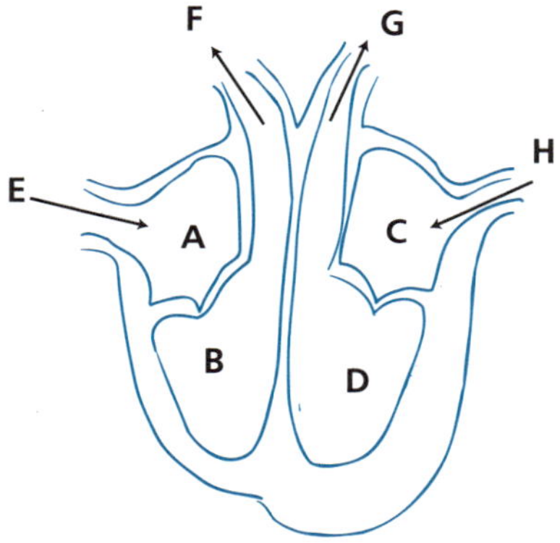

Part	Letter	Function
right atrium		
left atrium		
right ventricle		
left ventricle		
vena cava		
pulmonary vein		
aorta		
pulmonary artery		

Question 7

Complete the explanations in the boxes, then use arrows to show the route taken by an oxygen molecule from the lungs to the muscle cells in the arm. The first word of each process is supplied.

oxygen molecule in the lungs
pulmonary
left ventricle
aorta
capillaries

Question 8

On the diagram of an alveolus below, mark the gas flow with arrows and words.

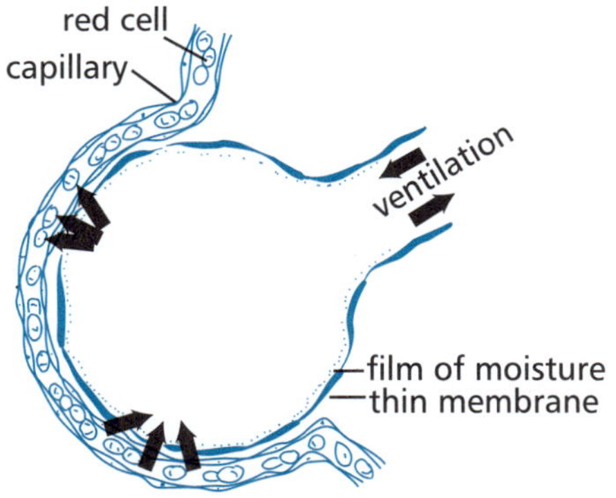

Question 9

On the diagram below label the parts indicated with arrows.

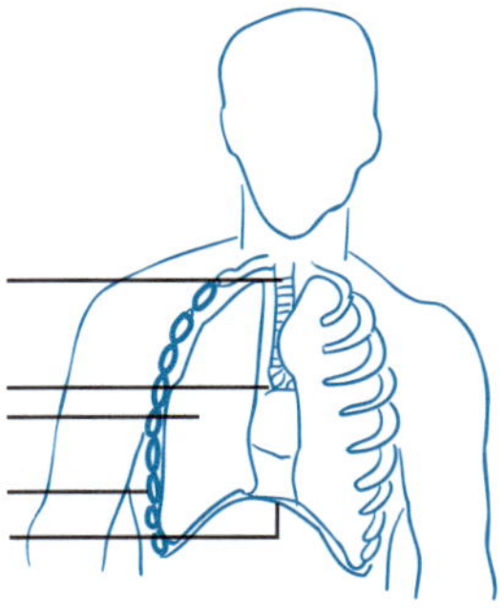

Question 10

When a peanut reacts with oxygen it produces energy. A group of students wanted to find out if a marshmallow also contains food energy.

They set up the experiment as follows:

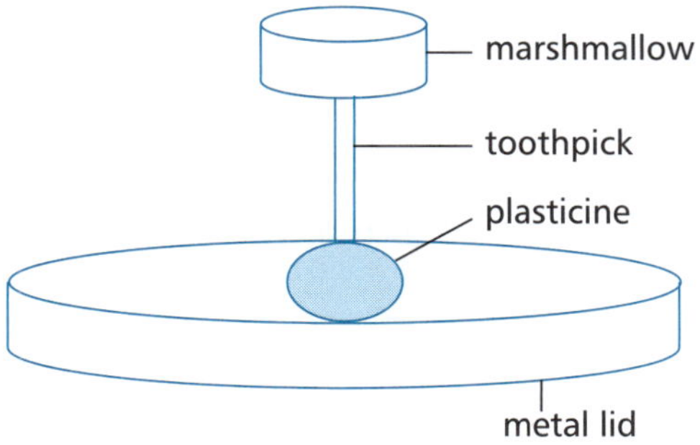

What will they have to do to the marshmallow to find the answer to their question?

____

____

Question 11

Study the diagram below and complete the table, matching the labels with the correct letters in the diagram. Write the function of each part in the third column of the table.

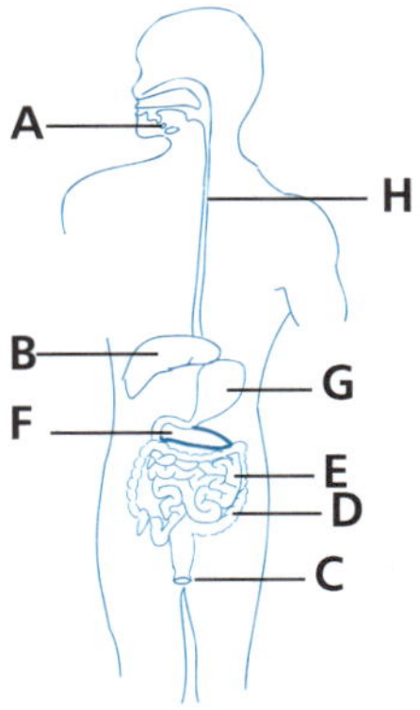

Part	Letter	Function
small intestine		
oesophagus		
liver		
stomach		
pancreas		
anus		
salivary glands		
large intestine		

Question 12

The table below shows how much of each substance is in the blood before and after going through the kidneys.

a Which substances are being filtered out?

____

b Where are the wastes stored?

____

c What is the name of the waste?

____

Substance	Plasma	Urine
water	910	960
proteins	80	0
urea	0.3	20
glucose	1	0
salt	7.0	9.5

Question 13

a Label the diagram with the words — urethra, ureter, kidney, urinary bladder.

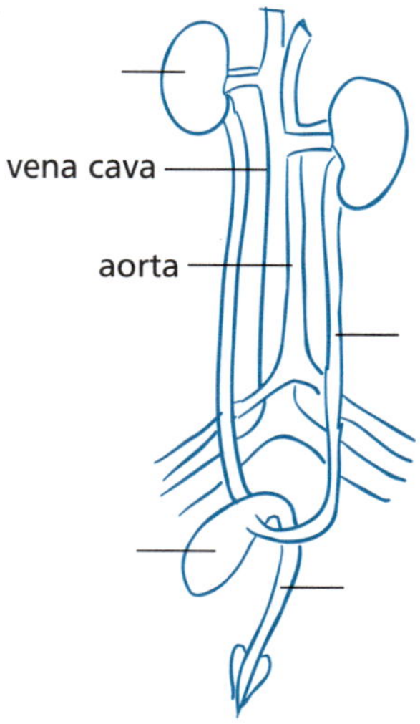

b Why do a major vein and artery go to the kidneys?

____

____

Question 14

This drawing shows some of the systems in the human body.

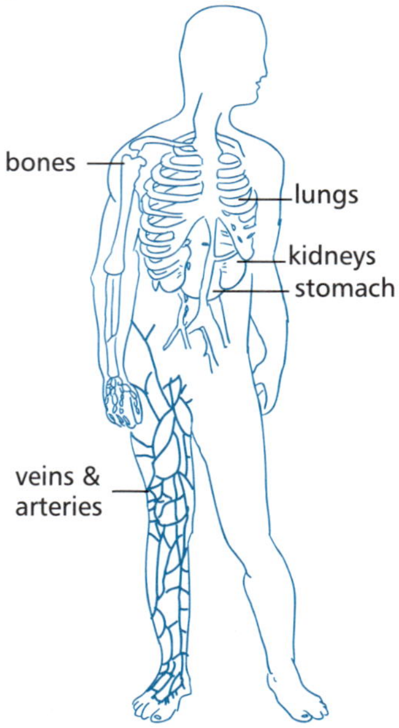

Identify which system each organ belongs to in the table below and label the diagram. Complete the table by adding the main function of each.

System	Organ	Function
respiratory		
digestive		
excretory		
skeletal		
circulatory		

Question 15

Put the following terms into the correct order — from smallest to largest — and match them with the right description:

organ system, cells, organ, tissue.

Descriptions:

a Collections of basic elements.

b Collection of organs performing a main body function.

c Group of similar cells specialised for one simple job.

d Collection of different cells grouped together to perform a particular function.

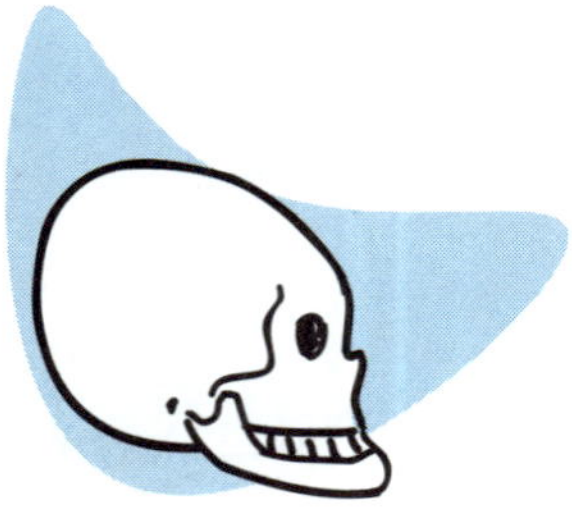

Revision questions

Question 1

List three things that the human body excretes and the organs that are responsible.

____

____

____

Question 2

What is the function of:

a red blood cells?

____

b white blood cells?

____

c platelets?

____

d plasma?

____

Question 3

Add the missing words to the sentences below.

When you breathe in your ____________ moves down. This ____________ the volume of the lungs and ____________ the pressure. So air rushes into the lungs. When you breathe out the ____________ moves up. This ____________ the volume of the lungs and ____________ the pressure, so air rushes out.

Question 4

a What is gas exchange?

____

____

b Where does gas exchange occur in humans?

____

____

c Why is gas exchange necessary?

____

____

Question 5

Match the words on the left with the definitions on the right.

Respiration	blood flow
Nutrition	expulsion of waste
Excretion	provides support and an anchor for muscles
Circulation	chemical reaction to breakdown food
Skeleton	obtaining raw materials to sustain life

Question 6

The following are diagrams of various types of joints. Label them with the correct term from the list below:

fixed joints, slightly moveable joints, synovial joints

A

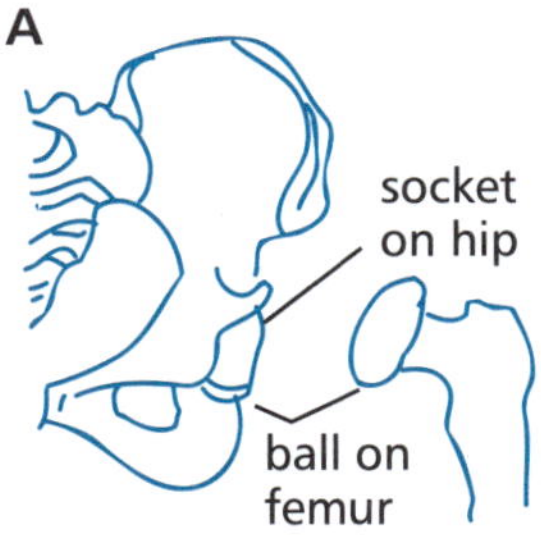

C

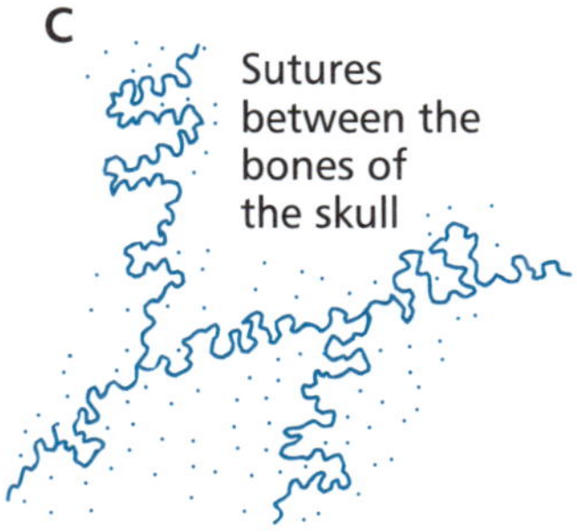

B

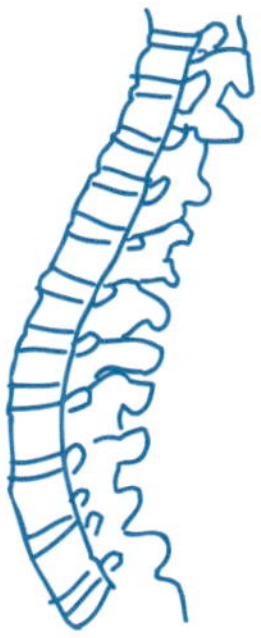

D

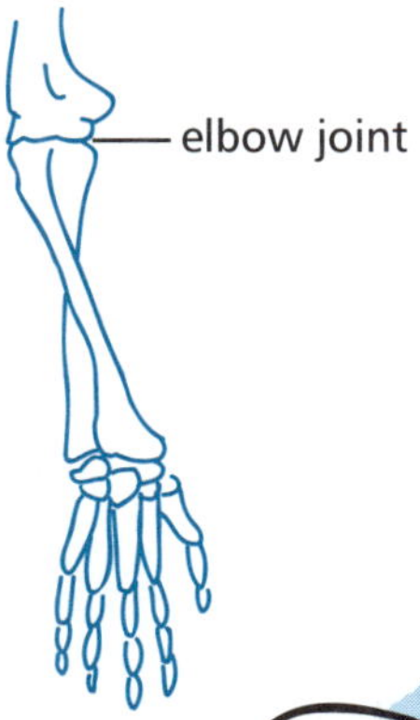

Question 7

Define the following terms:

a trachea

____

____

b bronchi

____

____

c alveolus

____

____

d lung

____

____

Question 8

Name the following bones:

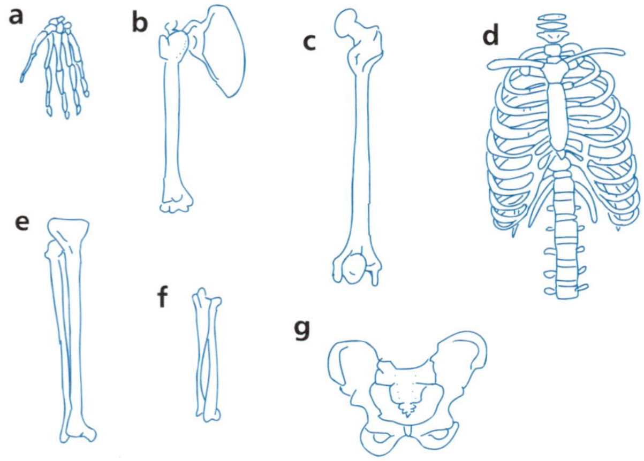

Question 9

The following diagrams are the limbs of mammals. Match the limb with the correct species:

dog, man, elephant, horse

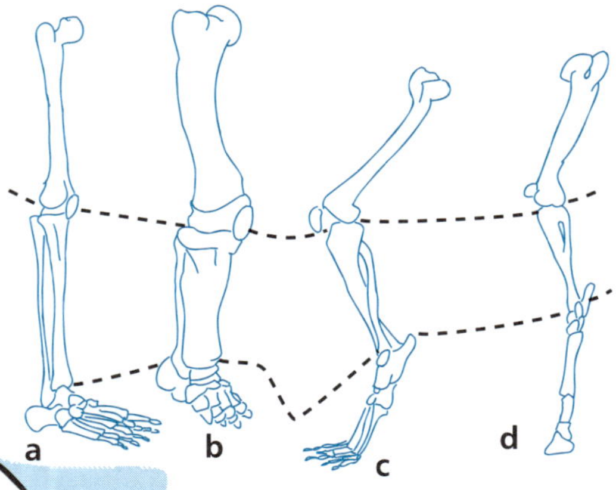

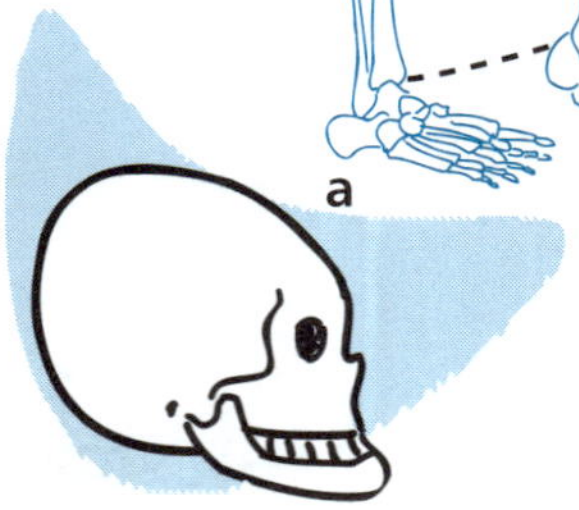

Question 10

a What is the main function of the excretory system in animals?

____

b What are three wastes that animals need to get rid of?

____

____

____

c Vertebrates live in all sorts of habitats. The kidney plays a key role in helping animals adjust to their surroundings. What are its two main functions?

____

____

d Explain how the function of the kidneys of a mammal living in the desert would differ from that of an animal living in the rainforest?

____

____

____

____

Question 11

a Name the human body systems sketched below.

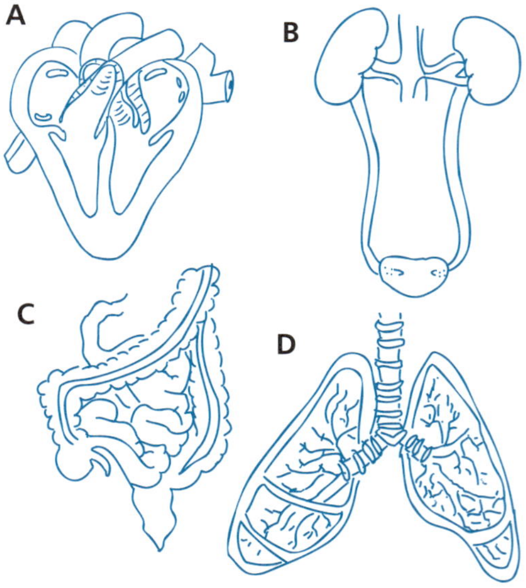

b List the main organs associated with each system.

A ______________________________

B ______________________________

C ______________________________

D ______________________________

Question 12

Study the structure of the following blood vessels and then label them with the terms:

artery, capillary, vein

a

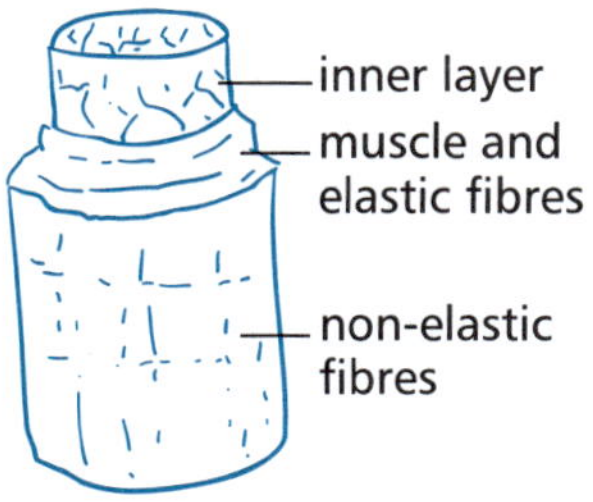

b

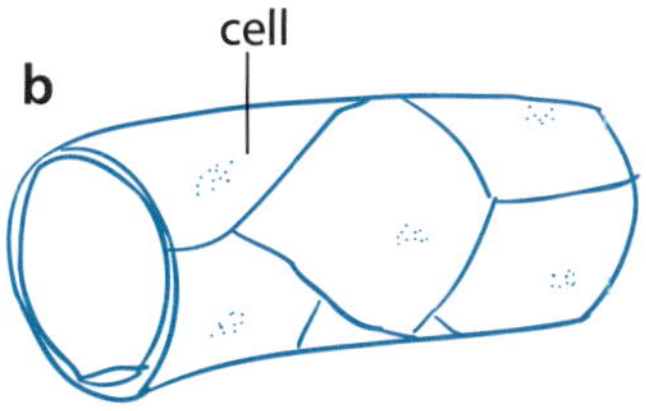

c

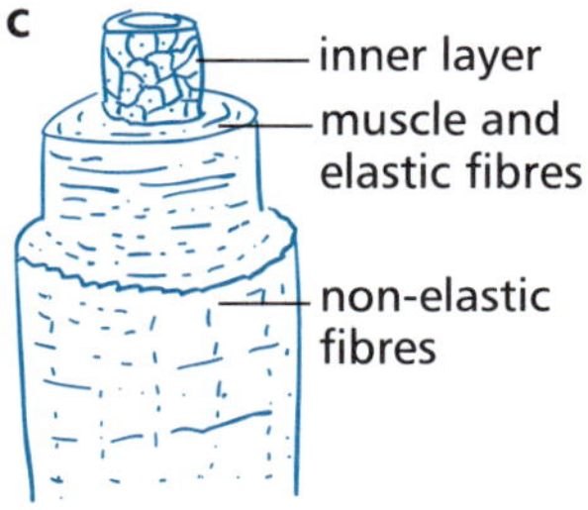

Question 13

Complete the following sentences.

In muscles, glucose and __________ pass out of the blood and into the cells. The cells use oxygen to break down the glucose and release __________. The process is a chemical reaction called ________________. A waste product called __________ __________ is produced which is carried by the blood to the __________ where it passes back out to the air.

Question 14

a Why does you heart beat faster when you exercise?

____

____

b When the heart pumps the artery stretches. The stretches can be counted. What are they called?

____

____

c What happens to blood flow in extremely cold conditions?

____

____

____

Question 15

a Why do large multicellular animals need systems for transporting things?

____

____

____

b What are four main functions of a transport system in mammals?

____

____

____

____

Question 16

Study the diagram of the heart then answer the questions.

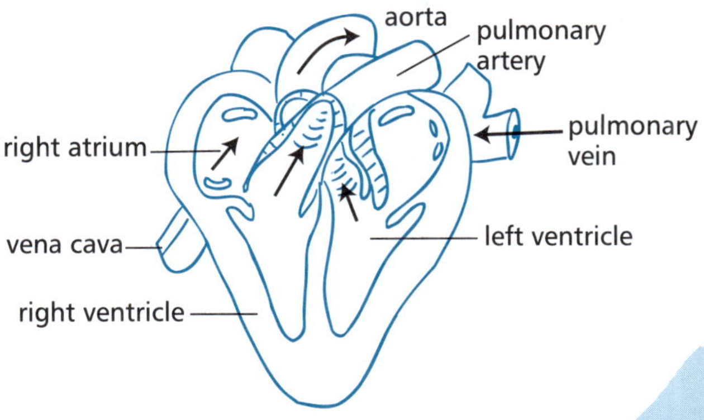

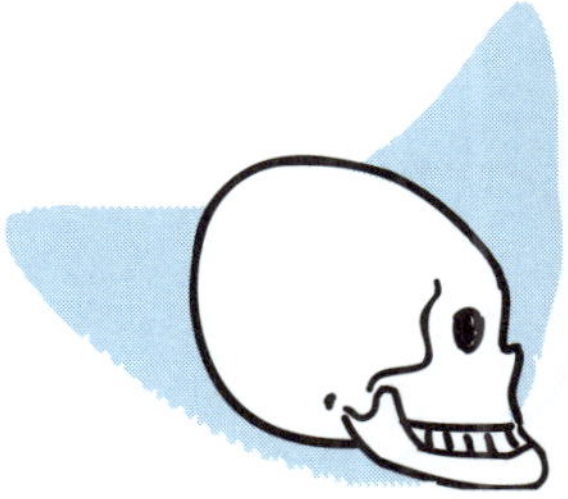

a Which side of the heart receives venous blood?

b Why is it necessary for the blood to circulate through the lungs?

c What is the main function difference between the pulmonary arteries and the pulmonary veins?

d Which carries blood away from the heart — arteries or veins?

e Could a vein be used to check a person's pulse?

Question 17

a The surface area of the lungs of an average adult is between 50 and 70 square metres. Why is it necessary to have such a large area available for gas exchange?

b What three important physical features does the gas exchange membrane in the lungs need to have to be efficient?

Question 18

The diagram below shows how food travels through the digestive system.

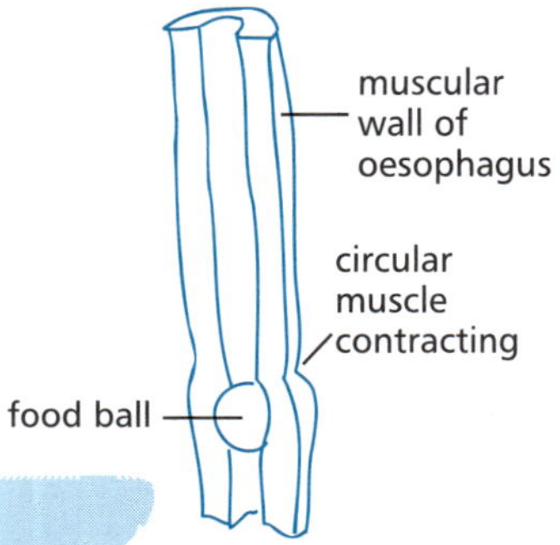

What is the name of this process?

Question 19

a What are the two main functions of the digestive system?

b Explain why animals have to digest their food.

c What is the main function of the stomach?

d What is the main function of the pyloric sphincter?

Question 20

Complete the sentences by filling in the missing words.

__________ is aided mechanically by the grinding of __________ and the churning of the __________. The rest occurs by chemicals called __________. __________ is produced in the salivary glands and __________ in the stomach. Other enzymes produced for digestion are renin, maltose and trypsin.

CHAPTER 3

CELLULAR REPRODUCTION

How do your cells help you grow?

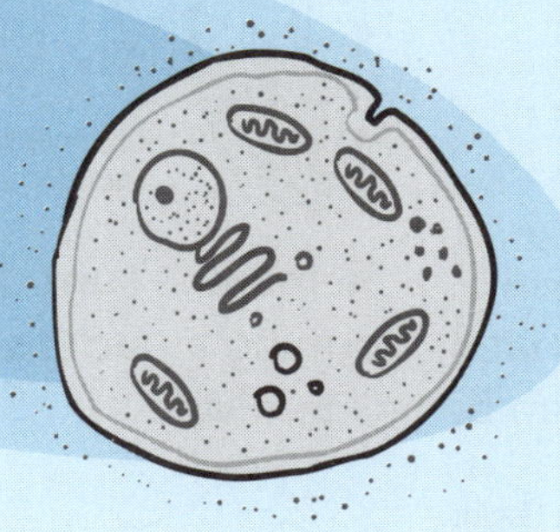

Living things make millions of new cells every day. They do this by cell division. When a cell divides its **chromosomes** separate in a process that makes sure each new cell receives the correct genetic material. Every cell has its own life cycle in which it reproduces itself or rests or divides and dies. Skin cells complete the cycle in 24 hours and others take much longer.

Cell division

As living things develop and mature they grow bigger but their cells stay almost exactly the same size. Growth is the result of cells dividing and increasing in number. There are two different ways that cells divide:

- **Mitosis** is the division of a cell nucleus to produce two identical cells.
- **Meiosis** is a form of cell division that produces the sex cells, which are all genetically different from each other.

The process of mitosis

Mitosis involves one division that produces two new cells with exactly the same genetic material as the original cell. Just before mitosis begins each chromosome copies itself and forms two new **chromatids**. During mitosis two chromatids pull apart to form two separate nuclei. Then the **cytoplasm** divides to form two identical cells.

Mitosis occurs in several stages. These are listed below.

- **Early prophase** — the chromosomes tighten up and condense and a network of tiny tubes called a spindle, begin to develop.
- **Late prophase** — the nuclear membrane disintegrates and the spindle begins to move the chromosomes.

Early prophase

Late prophase

- **Metaphase** — the chromosomes line up along the centre of the spindle.

Metaphase

- **Anaphase** — the chromatids (duplicate chromosomes) that make up each chromosome move apart and travel to opposite ends of the spindle.

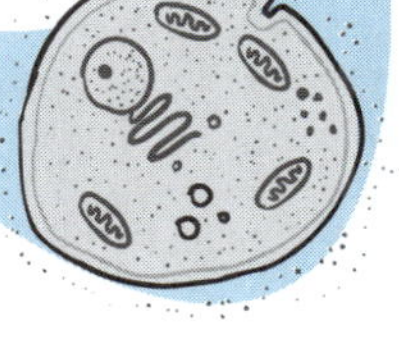

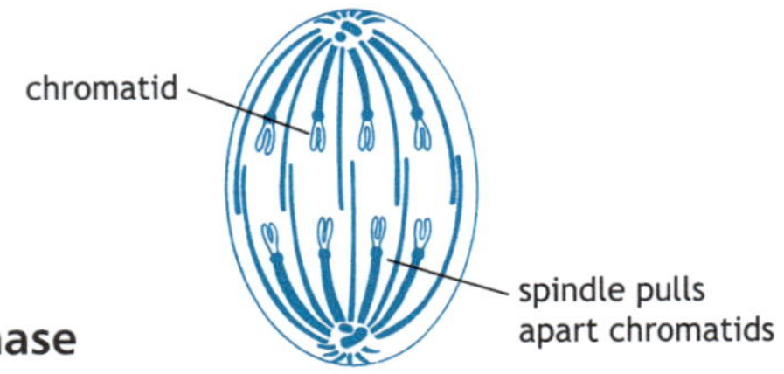

Anaphase

- **Telophase** — a membrane surrounds each set of chromatids, forming a nucleus. The cytoplasm begins to divide.

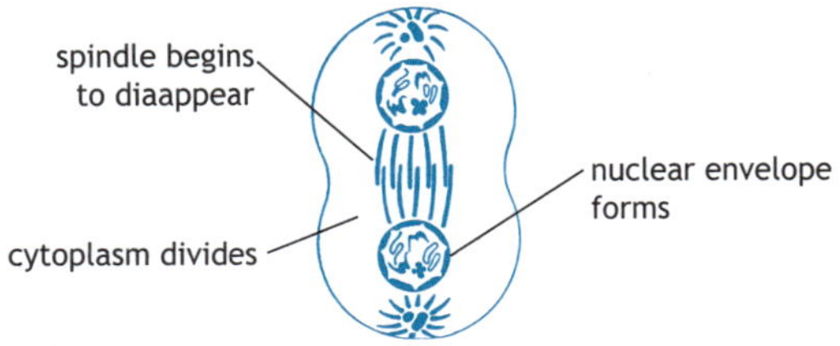

Telophase

- **Interphase** — once cell division is complete, the chromatids unwind. The two new cells have identical genetic material because the copies separate.

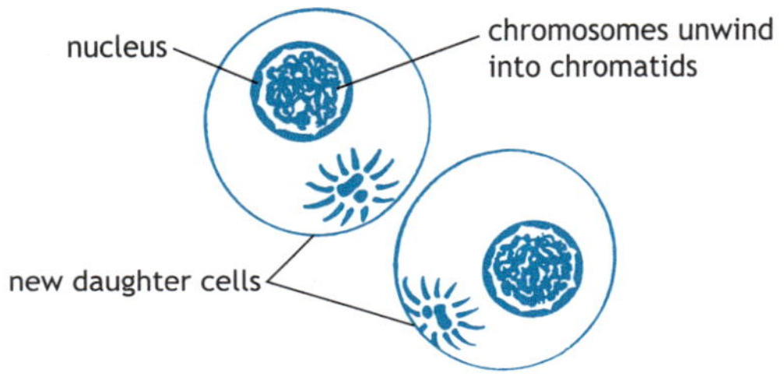

Interphase

Somatic cells make up almost all of the cells in living things. They divide only by mitosis and every new somatic cell has a double set of chromosomes. Somatic cells are therefore diploid. Somatic cells are not involved in reproduction, so they cannot pass on genes to the next generation. Human somatic cells have 23 pairs of chromosomes each.

The process of meiosis

Meiosis involves two divisions, one following the other, which produces four new cells with exactly half the genetic content of the original cell. During these divisions **genes** or **alleles** are mixed up so that each new haploid cell has a unique genetic make-up.

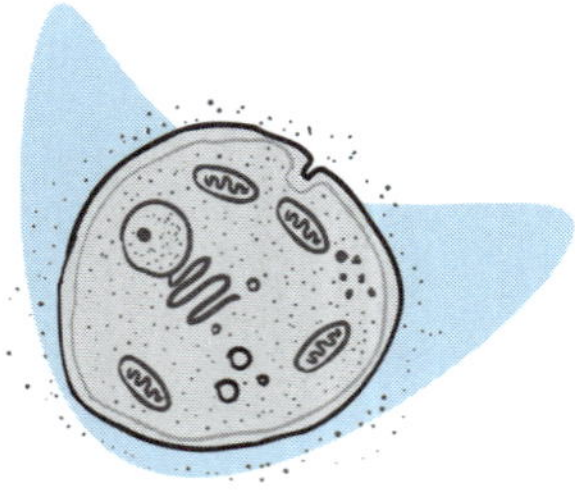

In Division I, pairs of **homologous** chromosomes separate to form two new cells with half the original number of chromosomes.

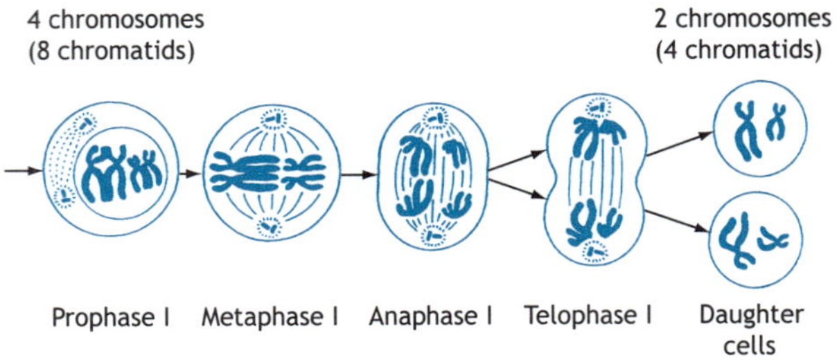

Meiosis — Division I

In Division II each chromosome replicates itself to form two sister chromatids. These chromatids separate and become chromosomes to form four cells with half of the original amount of genetic material each.

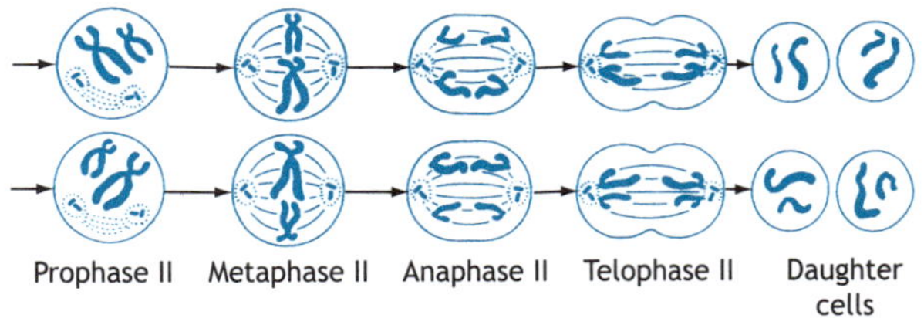

Meiosis — Division II

Sex cells, or gametes are formed by meiosis and are used in sexual reproduction. In humans the male sex cells are called sperm and are made by the testis. The female sex cells are called ovum and are formed by the ovary. Sex cells are not like the cells in the rest of the body as they only have half the genetic material — they are haploid. During sexual reproduction they pass their genes onto the next generation.

Human sex cells have 23 chromosomes each.

Why sex cells are not identical

When meiosis begins pairs of homologous chromosomes line up and form tetrads. Each chromosome has two chromatids — so a tetrad is a single unit of a set of four. The chromatids in each tetrad swap pieces with each other, recombine and then later pull apart. This **crossing over** gives each chromosome a new combination of genes.

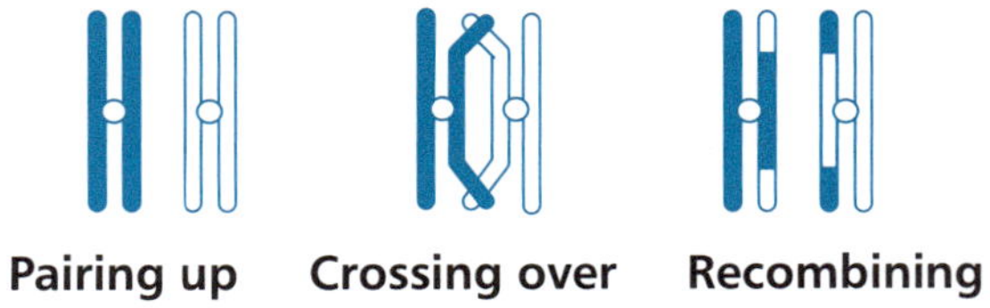

Growing and changing

Every human being starts life as two sex cells. When they join together they produce a single fertilised egg called a zygote.

Straight away the cell starts to split via mitosis into two, then four, then sixteen cells and so on.

After five days this ball of cells implants in the uterus wall and continues to grow. It forms a head, arms, legs and so on and grows thus:

- At 4 weeks the embryo is 1 cm long.
- At 8 weeks the foetus is 3 cm long.
- At 12 weeks it is 12 cm long.
- At 28 weeks it is 34 cm long.
- At 38 weeks the baby is 52 cm long and ready to be born.

At birth the baby is completely dependant but slowly learns to walk, talk, and control its body. The baby grows bigger and bigger through cell division. At puberty boys become more muscular, the penis thickens and they grow hair on their face, armpits and pubic area. Girls bodies change too, they grow body hair and they begin to menstruate. The testes begin to produce sperm and the ovaries to produce ovum by the process of meiosis.

Both males and females stop growing and cell division is mostly for replacing worn or damaged cells. As females grow much older they stop producing ovum and menstruating.

Things that can go wrong

The bodies of living things make new cells all the time by the process of mitosis. If cells divide too quickly or something causes them to be malformed, then disease such as cancer can result. Disease is a breakdown in the body's stable state.

- **Inherited diseases** are passed on from parents to children in the genes. Examples of inherited diseases are:
  - Sickle-cell anaemia — the red blood cells are the wrong shape and do not carry oxygen very well.
  - Cystic fibrosis — the pancreas does not function and enzymes for digestion must be injected.
  - Haemophilia — the inability of the blood to clot.
- **Nutritional diseases** are associated with an unbalanced diet. Anorexia nervosa (too little food), obesity (too much food) and anaemia (not enough iron) are all examples of nutritional diseases. Lack of vitamins can cause poor eyesight (vitamin A) and weak bones (vitamin D).
- **Physiological malfunction** are diseases caused by a malfunction of cells, tissues or organs. Examples are:
  - Heart disease — the reduction of blood flow caused by fatty layers in the blood vessels. Heart disease is related to smoking, poor eating habits, salt intake and lack of exercise.
  - Diabetes — there is not enough insulin produced to shift all the blood glucose into the cells. This disease can be related to obesity, too much sugar in the diet and lack of exercise.
  - Cancer — a group of diseases that result from uncontrolled cell growth. A tumor occurs when this 'jumble' of cells swells into a mass that serves no useful purpose. Substances called carcinogens are capable of causing cells to malfunction, for example, asbestos.

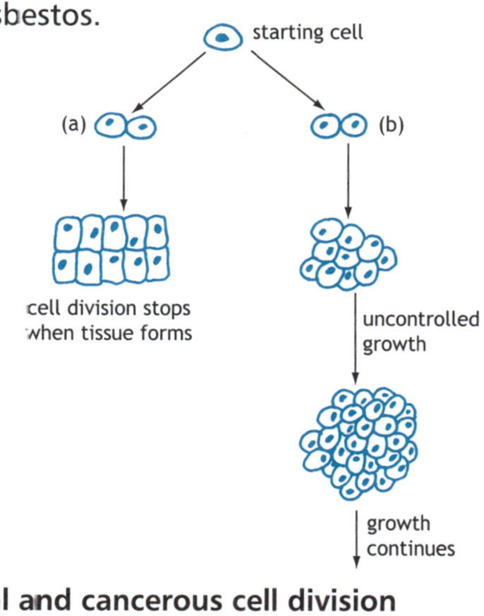

Normal and cancerous cell division

- **Environmental diseases**, triggered by our immediate surroundings can damage cells, for example: skin cancer (UV radiation), deafness (noise pollution), lung cancer (smoking) and diseases of drug abuse (alcohol).

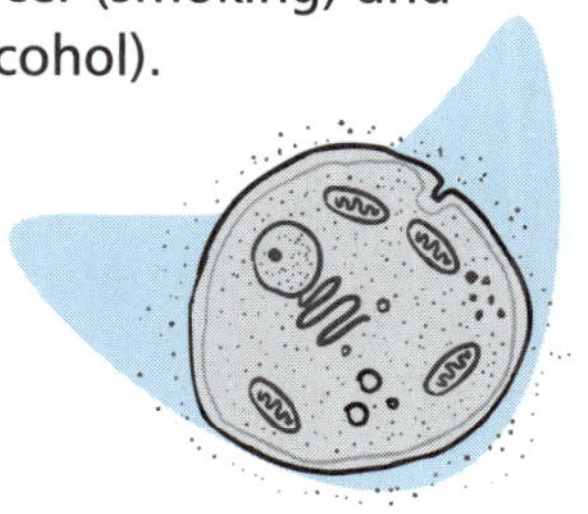

Keeping cells fit and healthy

- Physical activity — whether you are young or old exercising should be part of every day. It helps control body weight, strengthens bones and circulation.
- Good eating habits — a healthy diet with all food groups balanced.
- Be a non-smoker — smoking doubles the risk of cancer and heart disease.
- Reduce stress — the pressures of everyday life can cause stress that may result in headaches, stomach ulcers and lack of sleep.
- No alcohol — alcohol is a poison and your body tries to get rid of it. Too much alcohol damages the cells in the liver and slows the brain down.
- Drugs — legal drugs like aspirin (headaches) and antibiotics (bacterial infections) should always be taken exactly how the doctor says. Illegal drugs like ecstasy and cannabis also damage the cells in our bodies.

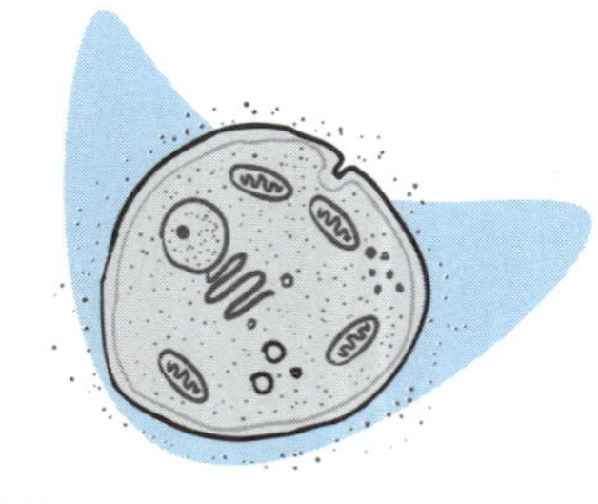

Glossary

Allele
One of two or more forms of the same gene.

Chromatid
One of two strands that are linked in a replicated chromosome.

Chromosomes
A structure that contains all or part of a cell's genetic material.

Cytoplasm
The contents of a cell, excluding the plasma membrane and nucleus.

Diploid
Two sets of chromosomes present in a cell (somatic cells).

Disease
A breakdown in the body's usual stable state.

Gene
A basic unit of heredity that carries the instructions needed to make a particular protein.

Growth
An increase in size due to increasing numbers of cells via cell division.

Haploid
One set of chromosomes in a cell (sex cells).

Homologous
Matching pairs of chromosomes in a diploid cell.

Mitosis
The division of a cell nucleus to produce two identical cells.

Meiosis
A form of cell division that produces the sex cells.

Sex cell
A cell involved in sexual reproduction.

Somatic cell
A cell that is not involved in sexual reproduction.

Science skills

Question 1

Match the following stages of mitosis with the correct diagram.

telophase, metaphase, early prophase

A

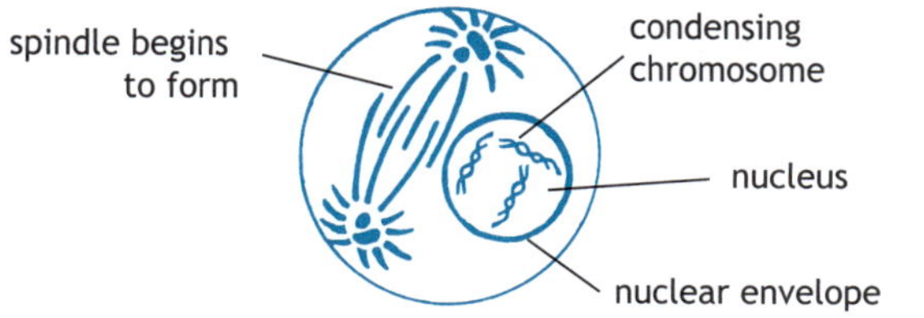

B

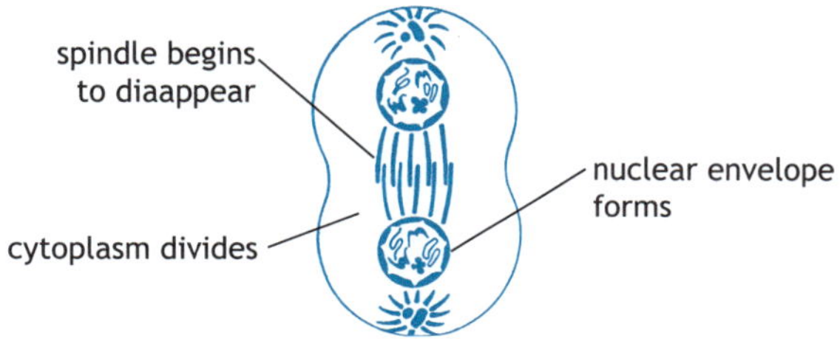

C

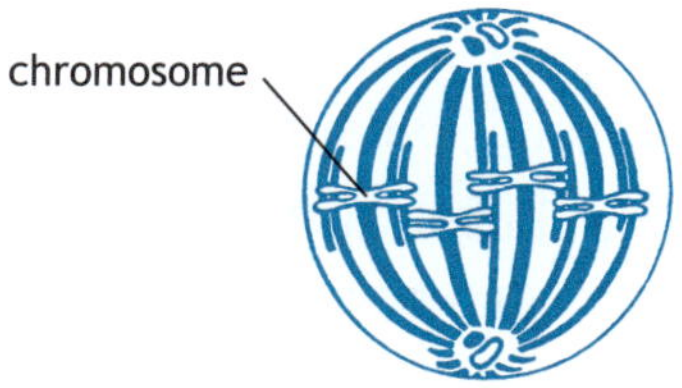

Question 2

The following diagrams represent the first and second divisions in the process of meiosis. Which is the first division and which is the second?

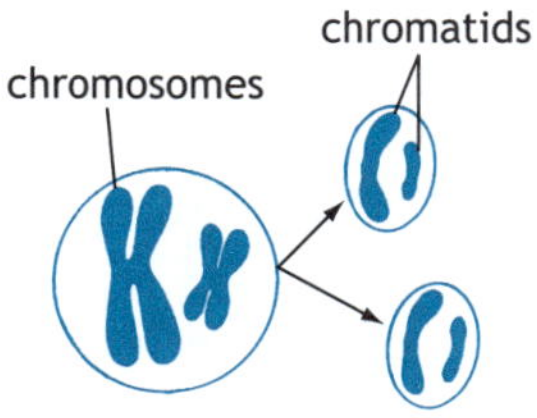

A

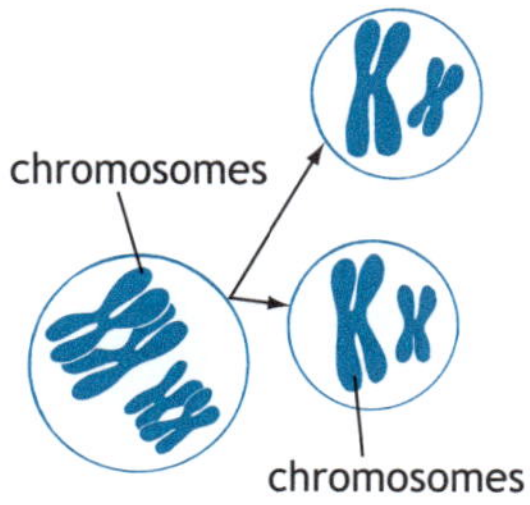

B

Question 3

Use the table below to compare mitosis and meiosis. Mark the relevant box with a tick.

	Chromosomes separate	Chromatids separate
Mitosis		
Meiosis (Division I)		
Meiosis (Division II)		

Question 4

Study the graph below and then answer the questions.

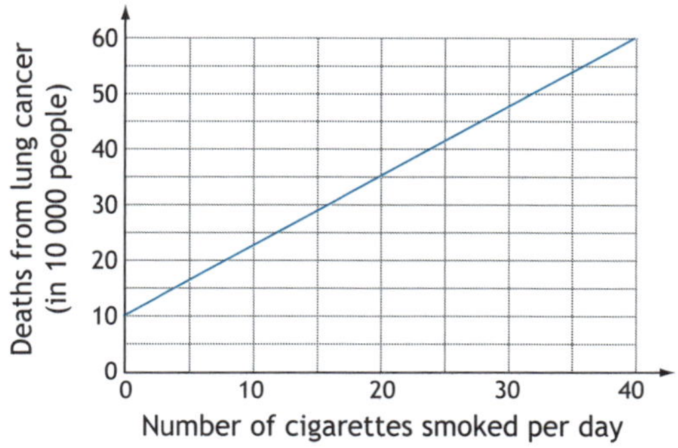

a How many people who smoke 30 cigarettes a day die of lung cancer?

____

b How many people who don't smoke die of lung cancer?

____

Question 5

Complete the diagram to show how nutrient deficiencies can damage the body.

Health problem	Cause
obesity	
poor eyesight	
anaemia	
weak bones	
anorexia nervosa	

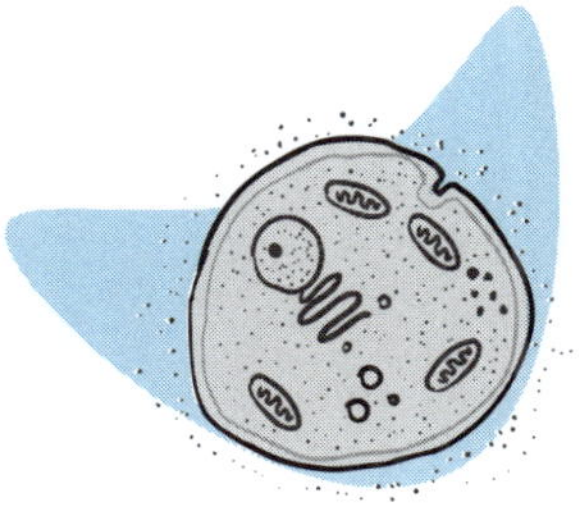

Question 6

Label the two processes of cell division drawn below.

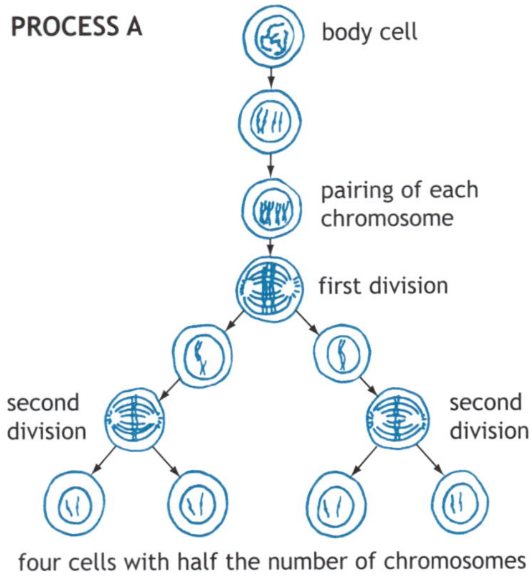

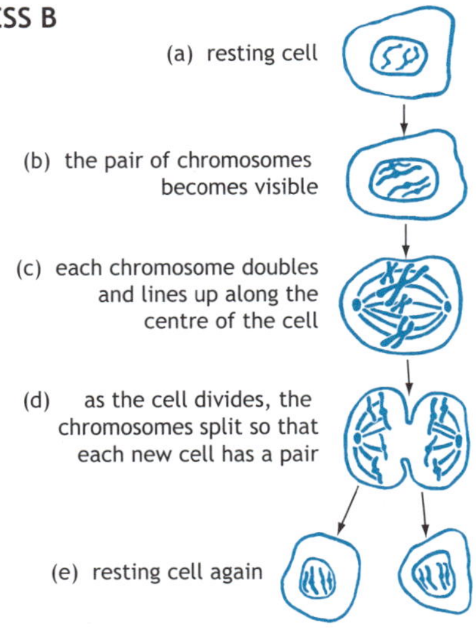

Question 7

Complete the box by writing in what causes the following non-infectious diseases.

Disease	Cause
sickle-cell anaemia	
diabetes	
deafness	
obesity	
lung cancer	

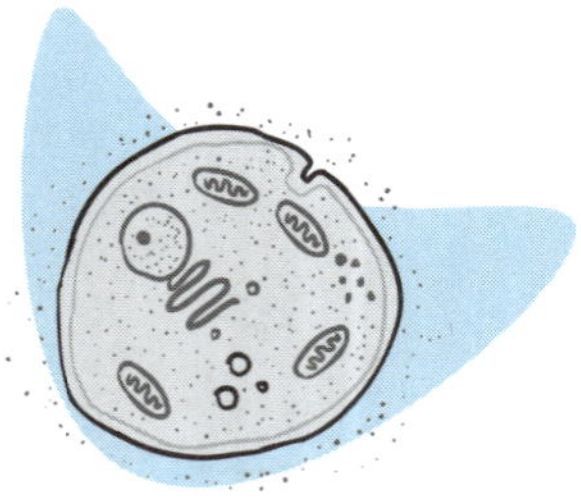

Question 8

The following table shows the amount of alcohol in an adult's blood after one glass of wine.

Time (hours)	Amount of alcohol in the blood (gms/litre)
0	8
1	6
2	4
3	2
4	0
5	0

a How long did it take for the liver to rid the body of alcohol?

b Estimate how long would it take to rid the body of two glasses of wine?

c How many grams of alcohol does the liver process every hour?

Question 9

If a malfunction occurs in a zygote then most body cells will carry the mistake. This is not so if the mistake occurs in a somatic cell.

a The two tadpoles below suffered some malfunction in certain body cells (marked with an 'x'. Which cell (A or B) is the somatic cell?

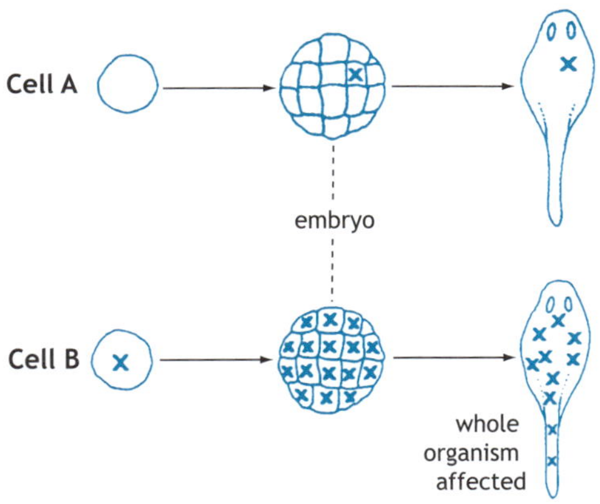

b What is the name for cell B?

Question 10

As children grow into puberty their bodies change. Fill in the table below with the sorts of changes that happen to boys and girls at puberty. One must be the same for boys and girls, two must be different and one is for boys only.

Girls	Boys

Question 11

Complete the table below to show what happens to a fertilised egg.

Weeks since fertilisation	Size (cm)	Name given to the stage of development
4		
8		
12		
38		

Question 12

The following compares somatic and sex cells. Tick the correct box to indicate their characteristics.

Characteristic	Somatic cells	Sex cells
they make up almost all of the body cells		
they divide by meiosis		
23 chromosomes		
diploid		
not involved in reproduction		
pass on genes		

Revision questions

Question 1

In ________, halving the number of ____________ ensures that the fertilised egg has the ______ number as the parents. _________ cells always have the same number of chromosomes as the parents.

Question 2

Answer true or false to the following statements:

a Alcohol is a drug.

b Most alcohol is removed from the body by the liver.

c Alcohol makes people more wide awake.

d Drinking a lot of alcohol can damage ones health.

Question 3

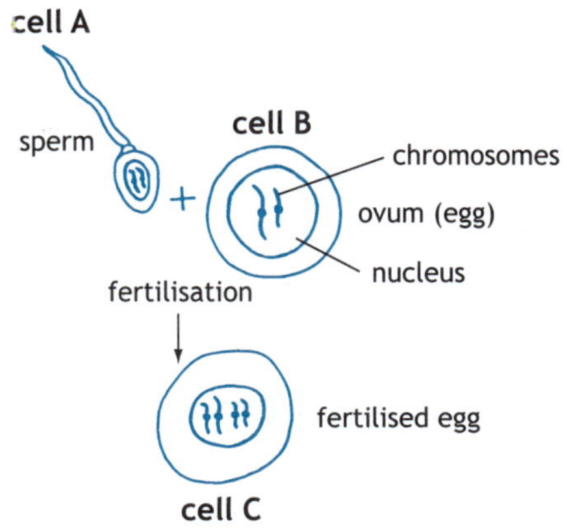

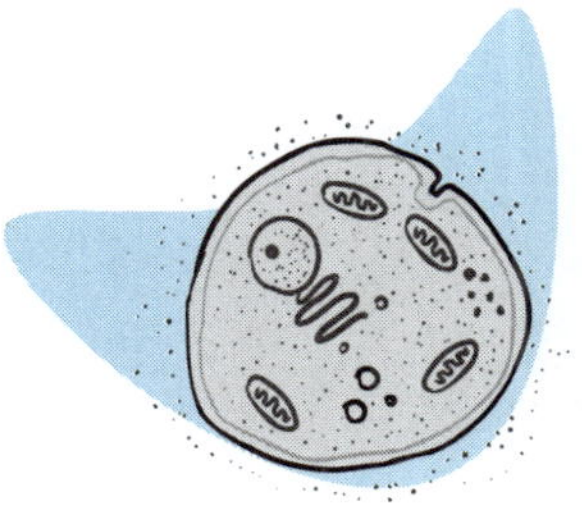

a What is the name of the process that produces cells A and B?

____

b What is the name of the process that produces cell C?

____

Question 4

Fill in the missing words:

Physical exercise is good for you because it helps your ______ pump blood around your body. Blood carries ______ and food to the cells and carries ______ and ______ ______ away. Activity also helps to control ______ ______ and ______ bones.

Question 5

a Why is it important that sun cream has UV protection?

____

b Could someone catch skin cancer from another person?

____

c What is skin cancer?

____

Question 6

Give one example of each of the following non-infectious diseases.

Disease	Example
inherited	
nutritional	
physiological	
environmental	

Question 7

Arrange the following five mitotic stages into the correct order.

anaphase, interphase, telophase, prophase, metaphase

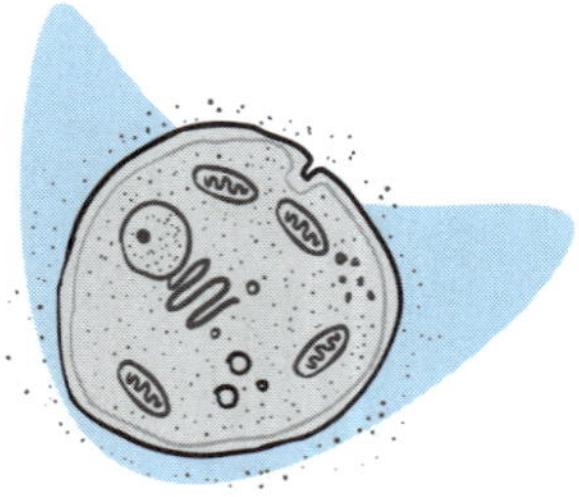

Question 8

Add the correct words to the following sentences.

______ ______ is the swapping of ______ between homologous chromosomes.

This provides each chromosome with a new combination of ______ from each ______.

Question 9

List six lifestyle choices that will help to keep cells fit and healthy.

1. ______________________________
2. ______________________________
3. ______________________________
4. ______________________________
5. ______________________________
6. ______________________________

Question 10

Label the diagram below with the following terms: nucleus, nuclear membrane, spindle, thickening chromosomes.

Question 11

Name the following two stages of mitosis.

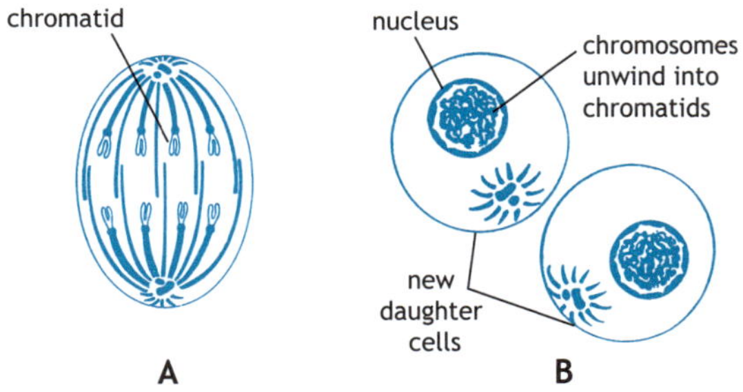

Question 12

Write the labels supplied below in the correct box on the diagram of meiosis given below.

- The cell has 4 chromosomes, two long and two short.
- First Division.
- The cell splits into two.
- Each chromosome replicates itself to produce two chromatids.
- Chromosomes separate and move to opposite ends of the cell.
- Second division.
- The chromatids line up in the centre of the cell.
- There is half the original number of chromosomes in each of the four cells.
- Chromatids separate and the cell splits again.

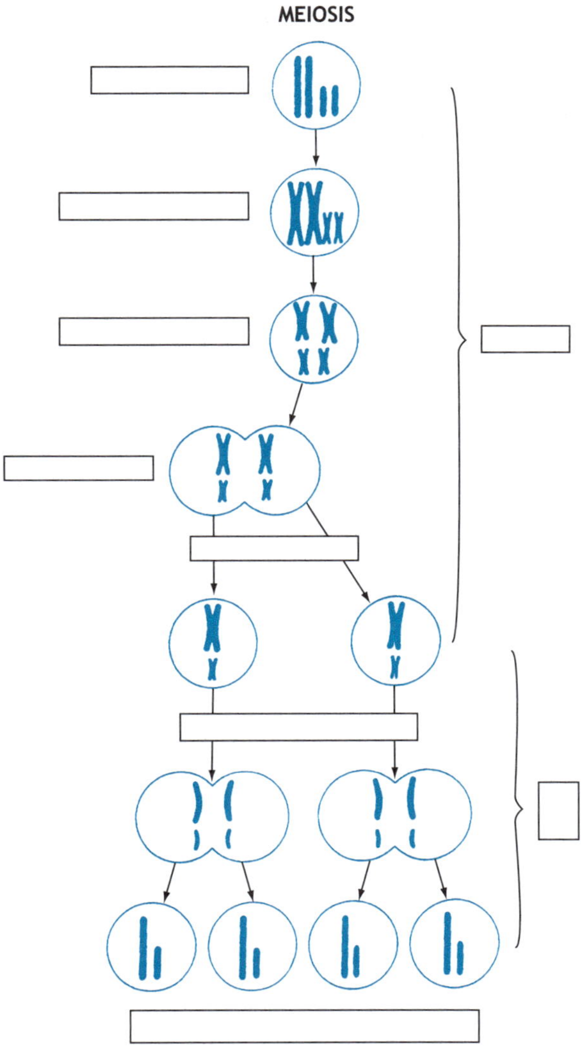

Question 13

a What are the two main ways that cells divide?

b Which of these forms is involved with normal growth?

c What happens to the number of chromosomes during mitosis?

d Where are chromosomes located in a cell?

e Draw a simple model to show the process of mitosis.

Question 14

a How do normal cells differ from cancer cells?

b What is a tumour?

c What is a carcinogen?

d What is an example of a carcinogen?

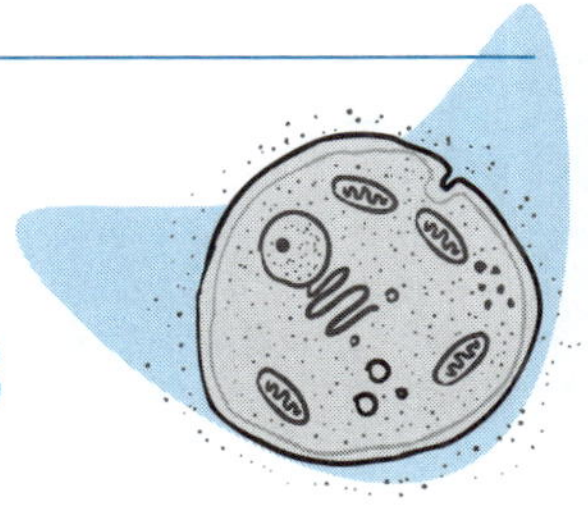

Question 15

In which of the following would you expect to find cells in the process of mitosis?

a The cheek cells of a person.

b The ovary of a mature female mammal.

c Some fertilised frog eggs.

d The skin of a person.

Question 16

Which of the following statements best describes the role of meiosis in vertebrate reproduction?

a Four daughter cells are produced

b Haploid gametes are produced to ensure sexual reproduction.

c Haploid gametes ensure an increase in species variation.

d Meiotic divisions eliminate undesirable genotypes.

Question 17

The number of chromosomes in a normal human diploid cell is:

a 23

b 46

c 44

d 26

Question 18

In which phase in mitosis do the chromosomes align along the centre of the spindle?

a prophase

b anaphase

c metaphase

d telophase

Question 19

Which of the following are illegal drugs?

a alcohol

b antibiotics

c cannabis

d aspirin

Question 20

_________ is a condition that exists when there is not enough ________ produced to shift all the blood glucose into the cells. This disease is related to __________ , too much ________ in the diet and lack of __________.

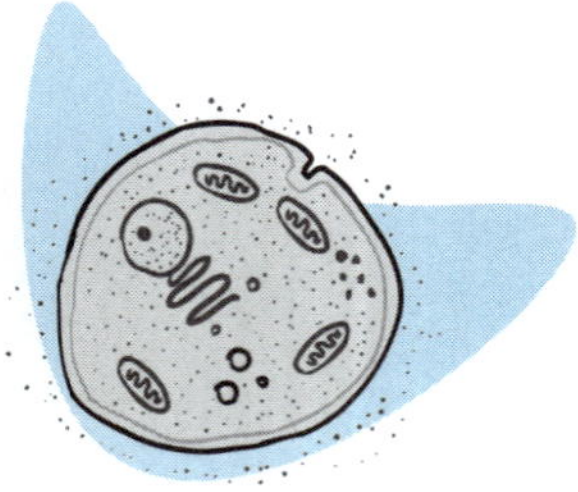

CHAPTER 4

MATTER

What is punching you in the nose right now?

What's matter?

Everything in the world can be organised into two groups:

1 **Matter** is collections of different kinds of atoms which take up space and can be weighed (eg cow, air, truck, ice-block).

2 **Energy** makes things happen and it has many forms (eg sound, light, heat etc).

Albert Einstein

This famous scientist said that matter and energy are interchangeable — if something is not matter, then it is energy and visa versa. Matter can be changed into energy and energy into matter. His equation for this idea is $E = mc^2$ where E = energy, m = mass and c = the speed of light. Since c is a very large number then a large amount of energy is formed from a small amount of matter.

Molecules

Matter is made up of clumps of tiny particles with empty spaces in between. The smallest clump of a substance that can exist by itself is called a molecule.

Atoms

This clump of particles can be broken up into even tinier bits called atoms. These are the very smallest pieces of a substance. There are about 1,000,000,000,000,000,000,000 atoms in a drop of water.

Where did matter come from?

The universe is a collection of atoms and the spaces between them. The same atoms have been here since the universe began and will be here until it ends. Our Earth is just a whole lot of atoms grouped together into the things we know (hamburgers, plants, speedboats, horses etc).

So matter is made of atoms, joined together to form molecules. The molecules join together to make matter.

Can we destroy matter?

If any new material appears in our lives it is just a group of the old atoms arranged in a different way. Matter cannot be created and it cannot be destroyed but its atoms can be rearranged.

Can matter be weighed?

Mass and weight

The amount of matter (atoms) in a substance is called its **mass**. Since matter is just lots of atoms, it can be weighed. How much something weighs depends on how much gravity is pulling on a substance (matter). How much gravity is pulling on an object depends on its mass.

Density

If the atoms in a material are packed very tightly together, the material will have more mass (a greater number of atoms) and therefore weigh more.

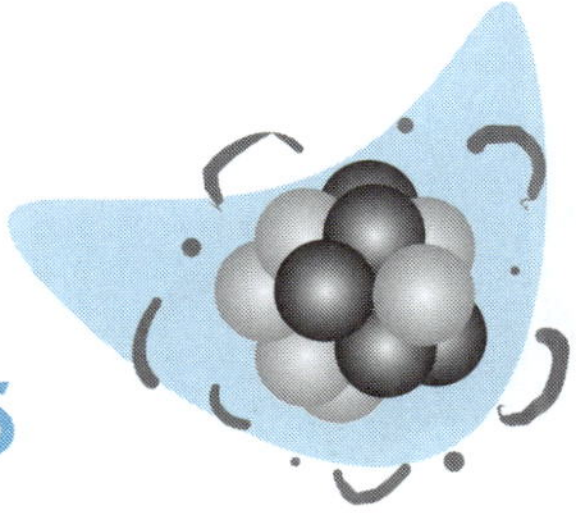

These materials are said to be very dense, for example lead. Materials such as marshmallows don't have many atoms but they do have a lot of air between the atoms, so they don't weigh much and are less dense. Gases are even less dense, and do not weigh much at all.

How heavy something is depends on its size and its density.

Measuring matter

- **Weight** — the weight of something depends on its mass and how much gravity is pulling on it. A sugar cube on Earth has the same number of atoms as a sugar cube on the moon but it will weigh less on the moon because there is less gravity.
- **Mass** is easy to work out — just put the object on some scales and measure how many grams it weighs.
- **Volume** can be measured by dropping the object into water and measuring the volume of water it displaces, or pushes aside (1millilitre = 1cubic centimetre).
- **Density** can be calculated by dividing the mass (grams) by the volume.

$$\text{Density (g /mL)} = \frac{\text{Mass (g)}}{\text{Volume (mL)}}$$

States of matter

On Earth matter can be a solid, a liquid or a gas. The difference between these groups is how closely packed together their molecules are. How closely packed the molecules are depends on how much they are moving. How much they are moving depends on how much energy they have.

All three 'states' have mass, volume and density.

Solids, liquids and gases

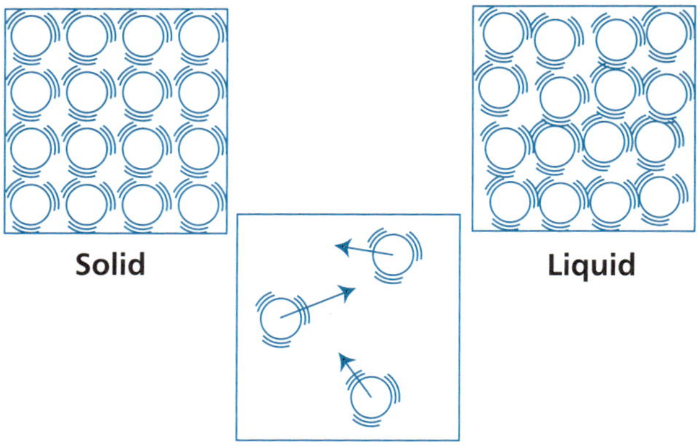

- **Solid**
  In a **solid** the atoms are packed close together in a pattern and vibrate on the spot, so the material maintains its shape. There are practically no spaces between the atoms and its volume stays the same.
- **Liquid**
  When heat (energy) is added to a solid the atoms start to vibrate more and move further apart. At a temperature called the **melting point** the atoms vibrate enough to break out and start sliding past each other and the solid becomes a **liquid**. The atoms of a liquid have some space between them and still vibrate but they also move around slowly. A liquid maintains its volume (the amount of space it takes up) but takes the shape of the container.
- **Gas**
  When heat (energy) is applied to a liquid the atoms start to vibrate even more and start to break out of the liquid (**evaporate**). At a temperature called the **boiling point** the atoms throughout the liquid vibrate and the liquid changes completely to a gas. In a **gas** the atoms fly all around the container, bumping into each other and the container walls. There are lots of spaces in a gas. A gas has no fixed volume and therefore takes up the volume of whatever container it is in. Gas molecules have the most energy.

Scientists call the vibration of atoms **Brownian movement**.

Changing the state of matter

When matter changes state it is still made of the same molecules. Energy must be added to, or removed from, materials to change their state.

The atoms in solids are held tightly together so it takes a lot of energy to overcome these forces and change it to a liquid. Atoms in liquids are held loosely but energy must still be added to change its state to a gas.

Solids have strong links …
and have to be given **energy** to become a liquid

Changing state

If energy is removed from the atoms of a gas, they slow down and become a liquid (condensation). If energy is removed from the atoms of a liquid they slow down even more and return to a solid (freezing).

This idea about atoms moving is called **kinetic theory**.

Sublimation

A few solid substances, such as dry ice, mothballs, iodine crystals, ice in a frost-free refrigerator, change directly to a gas when they are heated. This change from solid to a gas is called **sublimation**.

Molecular forces

Diffusion

When atoms or molecules are moving around, as in liquids and gases, they spread out all through the container. It takes a long time for liquids to diffuse, because the atoms are moving slowly. Gases diffuse fast because the atoms are moving fast.

Surface tension

There is a strong force between molecules on the surface of a liquid. This is why insects can walk on water and why droplets of water hang together.

Cohesion

This is a force between molecules of the same substance. For example, cohesion holds the water molecules in raindrops together.

Adhesion

This is the force of attraction between different types of molecules. For example, adhesion holds paint to the wall.

Capillary action

Water in a narrow glass tube will rise because the adhesion (between the water and glass molecules) is greater than the cohesion (between water molecules).

The capillary action occurs when you use a towel to dry your face; in babies' nappies and paper towels; and when rain water soaks into the soil.

Properties of materials

- Solids, liquids and gases have different properties, ie they behave differently.
- The properties of a substance are the qualities that make it different from other substances (hardness, density, colour).
- The properties of a substance determine the ways that substances can be used. Liquids (water) and gases (gas) can come into houses through pipes, but solids cannot.

What are some properties of solids, liquids and gases?

- Liquids change shape to fit the container they are in. Lumps of solid keep the same shape but a gas will spread out to fill any space. Ice (solid) melts to water (liquid) and then to steam (gas).

- Solids and liquids are heavier and denser than gases. If a substance is less dense than water (eg oil), it will float on water. If it is denser (eg iron) it will sink.
- Gases can be squashed and then they can spring back (eg the air bag in a car). Because there are large spaces between the molecules they can be pushed together (**pressure**). It is very difficult to squash liquids and solids as there is not much space between their molecules.
- Solids and liquids stay the same volume unless we heat them up (expansion) or cool them down (contraction). Bridges are constructed in sections so that there is room for the metal or concrete to expand and contract depending on the weather.
- Gases also expand if heated, and if the volume is not increased pressure develops inside the container. This pressure can be used to do things.

Glossary

Atom
The smallest particle of a substance.

Boiling point
The temperature at which a liquid boils.

Brownian motion
The vibrating motion of particles.

Density
The mass (number of molecules) in a certain volume.

Energy
Energy makes things happen or causes changes.

Evaporate
When a liquid changes to a gas.

Gas
Substances that spread out to fill all the space they can. They have no fixed shape or volume. They can also be squeezed into a smaller volume.

Kinetic theory
This theory suggests moving particles.

Liquid
Substances that have a fixed volume but take the shape of the container.

Matter
Anything that takes up space and can be weighed.

Mass
How much matter there is in something.

Melting point
The temperature at which a solid changes to a liquid.

Molecule
The smallest part of a chemical compound.

Pressure
The combined force of all moving molecules within a substance as they hit anything in their way.

Solid
Substances that stay in a definite shape.

Sublimation
The process whereby a substance changes from a solid to a gas without becoming a liquid first.

Science skills

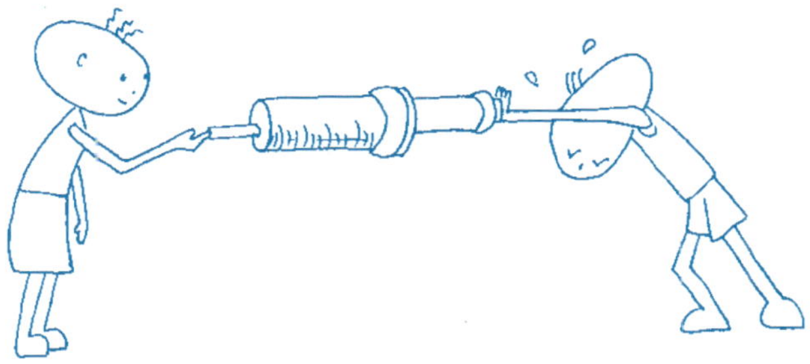

Question 1

Fill in the boxes with 'yes' or 'no'.

Property	Solid	Liquid	Gas
Takes up space			
Has weight			
Fixed volume			
Fixed shape			
Has strength			
May exist as crystals			
Can be compressed			
Density (high, medium, low)			

Question 2

Tick which group the object belongs to. Some are a bit tricky!

Object	Solid	Liquid	Gas
Glass			
Shaving foam			
Soft drink			
Perfume			
Syrup			
Chalk			
Cotton wool			
Air			
Rubber bands			

Question 3

Two students are doing an experiment on matter. They have a syringe full of air. They have pulled the plunger right out and put a finger over the opening.

a Can they squash the air inside the syringe?

b Explain your answer.

c Will the plunger move if they let it go? Why?

d If the syringe was full of milk would they be able to push the plunger in?

e If the syringe was full of salt would they be able to push the plunger in?

f If a gas is squeezed so that its volume is halved, what will happen to the number of particles per mL? What will happen to the density?

Question 4

Density measures the packing of particles.

$$\text{Density} = \frac{\text{mass}}{\text{volume}}$$

The density of steel is 7.9 g/cm^3, the density of water is 1 g/cm^3.

a Work out the density of the materials in the following diagram.

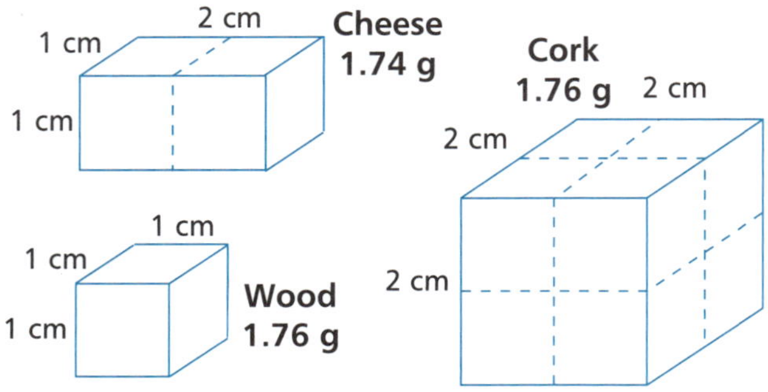

b Write down the materials in order of their densities, starting with the highest.

Question 5

Study the following table and answer the questions.

Gas	Density at normal temperature & pressure (g/litre)
air	1.3
hydrogen	0.09
helium	0.18
carbon dioxide	1.98

a What is the density of air?

b If water has a density of 1 g/mL then how many times denser is water than air?

c If 1 cubic metre is 1000 litres, what is the volume in litres of a garden shed measuring 2 m × 2 m × 3 m.

d What is the mass of air in the shed?

e Which gas would lift a balloon the fastest?

f Which gas would you find closest to the Earth?

Question 6

Match the diagrams with the correct state of the chocolate.

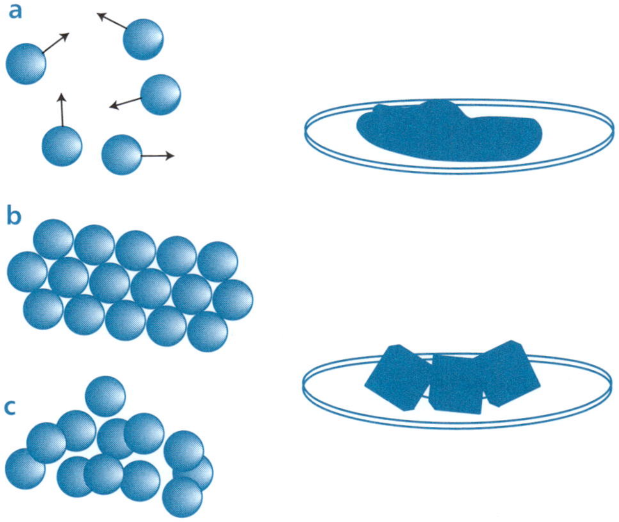

Question 7

When heat (energy) is added to a solid the particles vibrate faster.

a Look at the graph of heating ice and complete the table below.

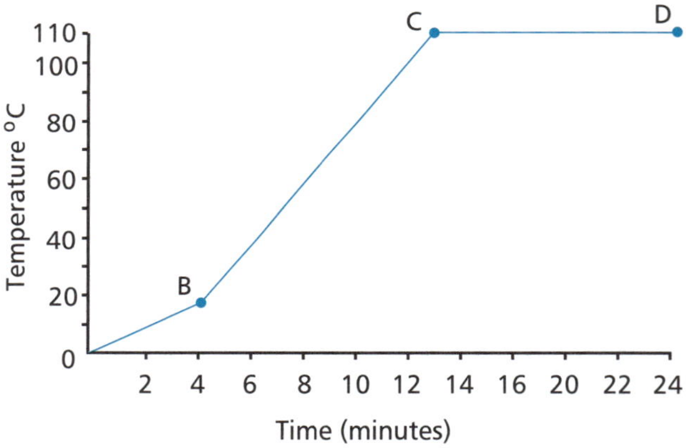

Graph of ice melting

	Part of graph		
	A to B	B to C	C to D
What is happening to the temperature?			
What is happening to the particles?			
Is it a solid, liquid or a gas?			

b What would you have to do to change this material back to a solid?

____

____

Question 8

a Put the following diagrams in order from the highest pressure to the lowest pressure.

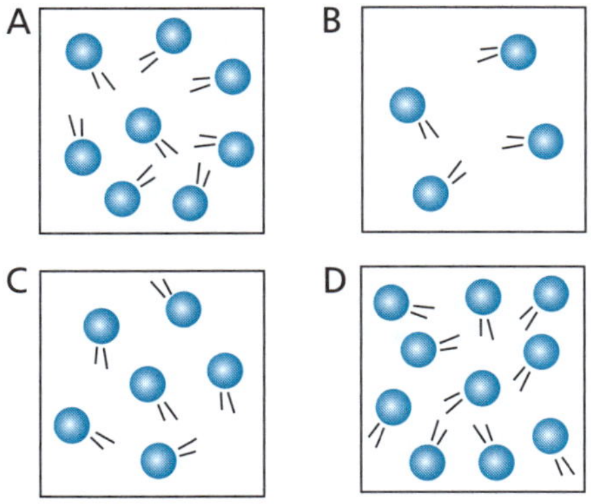

Highest ____________________ Lowest

b Explain your answer to **a**.

____

____

Question 9

A student heated a solid and recorded the temperature every 2 minutes. The data is recorded below.

Plot the data on a graph. Circle each point, then connect them with a line.

Time (minutes)	Temperature (°C)
0	20
2	30
4	40
6	45
8	50
10	55
12	60
14	65
16	80
18	90
20	100

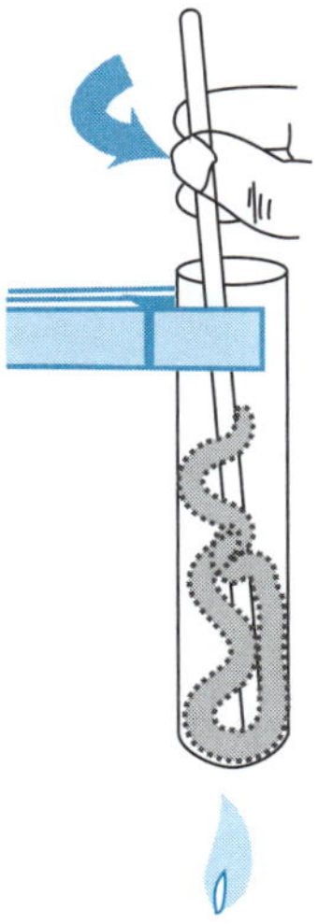

Question 10

Listed in the table below are several substances and their densities.

Substance	Density in g/mL
bone	1.8
clay	2.0
diamond	3.0
ice	0.9
slate	2.6
cardboard	0.7

Answer the questions below assuming that the volume of each substance is the same.

a Which one has the greatest density?

____

b Which one has the smallest number of molecules? ____________________

c Which one has the largest mass? ____________

d Which one weighs the most? ______________

e Which one has the most spaces between its molecules? ____________________

f If water has a density of 1g/mL, which substances will sink? ______________

g Why do icebergs float? ____________

h Which substance is twice as dense as ice?

____

Question 11

Label the columns with the correct state.
Particles are:

	a __________	b __________	c __________
packed	closely	spaced out	closely
arranged	regularly	at random	irregularly
moving	on the spot	very fast	past each other

Question 12

Students are working on an experiment.

They have stretched a balloon over the top of a small glass bottle and put the bottle into boiling water.

a What do you predict will happen to the balloon? Explain why.

b If the students put the bottle into iced water what do you predict will happen? Explain why.

c What use could you put this scientific principle to?

Question 13

a Does water have holes in it?

b Work out an experiment to prove this.

Question 1

Count how many molecules are in these bottles.

a Which bottle weights the most? Why?

b Which one has the smallest forces between its molecules?

Question 2

What is the name of the process that gets the perfume from the bottle to the nasal cavity.

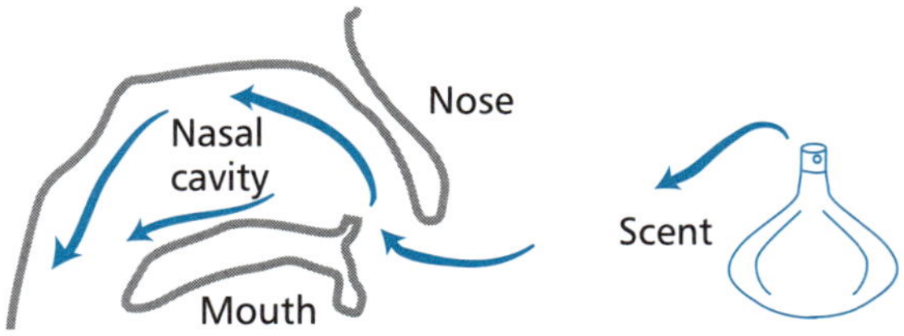

Movement of molecules

Question 3

A student set up the apparatus below and observed no evidence of liquid during the experiment, but a purple gas appeared.

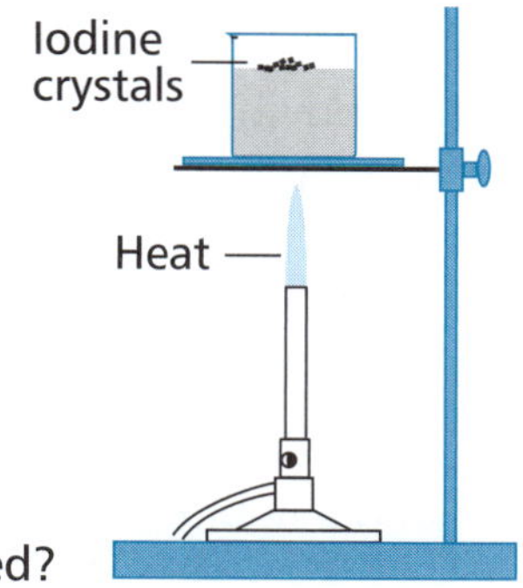

a What is this process called?

b How many changes of state occurred in this experiment?

c What would happen if you did this experiment using dry ice?

Question 4

Label the following pictures with the words melting, freezing, condensing, evaporating and boiling.

Question 5

If a person can run through air, why can't they run through a wooden fence?

Question 6

What happens to a gas if it is not confined inside a container?

Question 7

Here are two dogs.

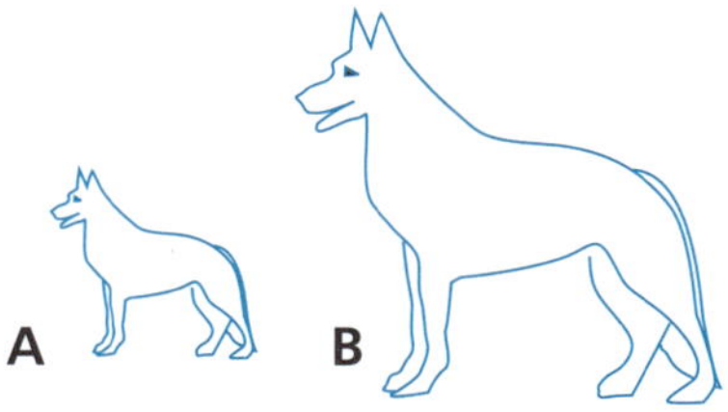

a Which one weighs the most on Earth?

b Which one weighs the most on the moon?

Question 8

a Which weighs the most: a kilogram of rice bubbles or a kilogram of butter?

b Which is the most dense?

c If a kilogram butter container is filled with rice bubbles and compared with a kilogram of butter which will weigh more?

d Why is it easier to throw a light rock further than a heavy rock?

Question 9

a In a swimming pool, who float better, fat people or skinny people?

b Who is less dense, boys or girls?

c How come submarines can float in water at any depth?

d If a freshly laid egg sinks in water, why does a rotten one float?

e Ships have a line on their hull called the Plimsoll line. Ships can only be loaded to this limit. If a ship is loaded to its limit in a very cold country what happens when it sails to a very hot country?

f Fresh water has a density of 1.0 g/mL. Why does sea water have a density of 1.03 g/mL?

g Why does popcorn pop?

Question 10

If a very narrow tube is dipped into some coloured water, the water will rise up the tube.

This is called capillary action. If the same tube was dipped in mercury what would happen? Why? (Note: Mercury has a density of 14 g/mL).

Question 11

Write down:

a one similarity between a liquid and a gas.

b one difference between the particles in a gas and a liquid?

Question 12

Fill in the missing words.

When you heat up a solid, you make the particles vibrate ________. The particles start to move around if they ________ strongly enough. The solid has changed into a ________. It has ________.

Question 13

a Where does the water go when wet clothing hangs on the line to dry?

b Do different materials take the same time to dry?

c What difference do you think spin drying will have?

d Do all parts of a garment dry at the same time?

Question 14

Predict which container of water will evaporate the fastest. Why?

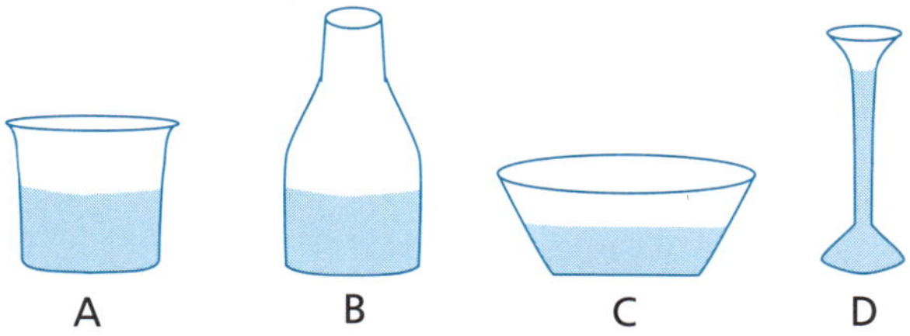

Question 15

Why do air conditioners in buildings and cars drip water?

Question 16

When we get out of the sea, we often feel colder than when we were swimming. Why?

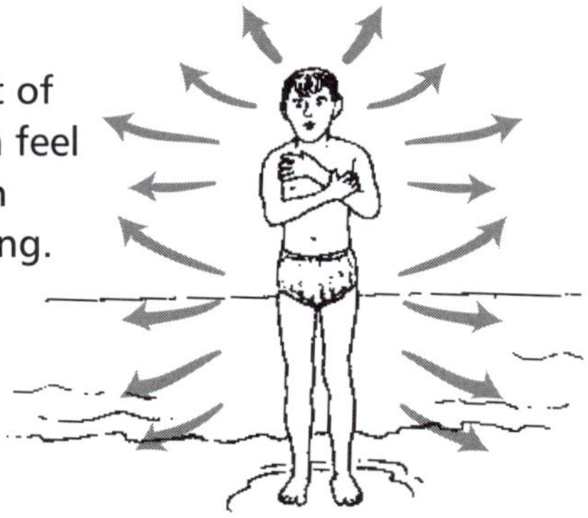

Question 17

An iron nail is 30 mm long at room temperature. When it has been heated it is 30.05 mm long. Explain what has happened.

Question 18

A student tied a cotton thread across a wire frame and dipped it into some soap.

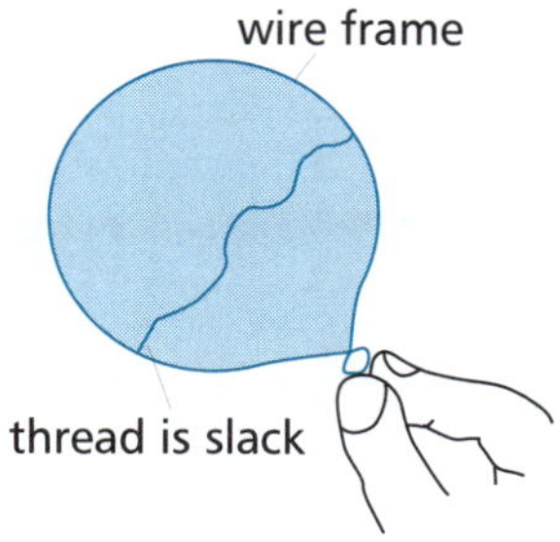

a What do you predict will happen if the student breaks the soap film on one side of the thread?

b What does this experiment show?

Question 19

The table below lists five changes of state. For each change, decide whether energy must be added or subtracted. Tick the right columns.

Change of state	Add energy	Subtract energy
a solid to liquid		
b liquid to gas		
c gas to liquid		
d liquid to solid		
e solid to gas		

Question 20

a The runways at airports are built of concrete slabs separated by thick bitumen dividers.

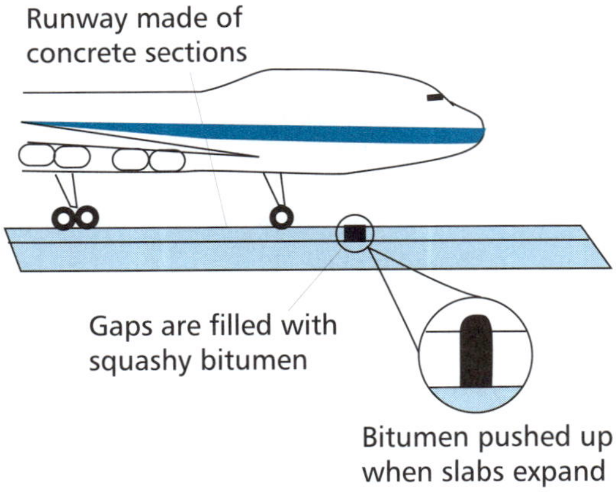

Explain why they are constructed like this.

____

____

____

b Telephone wires are put up so that they are slack.

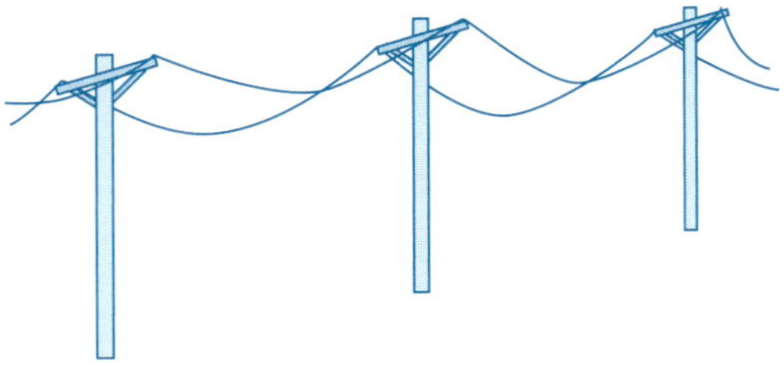

Explain why this is done.

____

____

____

Test A

Question 1

Complete the table using key words.

Vertebrate type	Description	Example
	has scales, breathes through gills, lays eggs	
	moist skin, lives on land & water, lays eggs	
	hair covers the body, young feed on mother's milk	
	dry scaly skin, eggs have tough leathery shell	
	wings, feathers, eggs have a hard shell	

Question 2

About 75% of all invertebrates belong to the arthropod group. There are many kinds of arthropod but they all have a skeleton on the outside and their legs are jointed.

Complete the table below with a description of each.

Group	Description	Example
crustaceans		crayfish
insects		fly
spiders		huntsman
lots of legs		millipede

Question 3

a Add the correct words.

Scientists sort living things into 5 main groups: ____________, ____________, ____________, ____________ and ____________. Animals without backbones are called ____________. If they have a backbone they are called ____________.

b Give an example of the following invertebrate groups.

Invertebrate Groups	Example
coelenterates	
round worms	
annelid worms	
molluscs	

Question 4

Use the key below to identify the tropical fish A, B, C, D, E. Note: Barbels are the projections from the mouth.

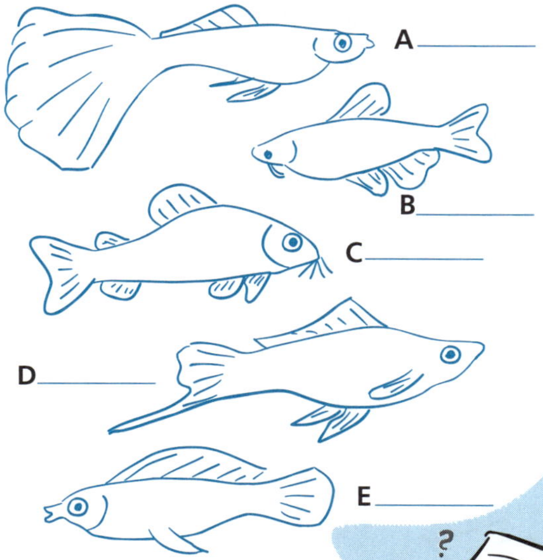

Description	Name/ instruction
1 with barbels without barbels	go to 2 go to 3
2 with 4 barbels with 2 barbels	Catfish Barb
3 with long point on tail fin without long point on tail fin	Swordtail go to 4
4 with large tail fin without large tail fin	Guppy Molly

Question 5

a Label the parts of the respiratory system on the diagram below.

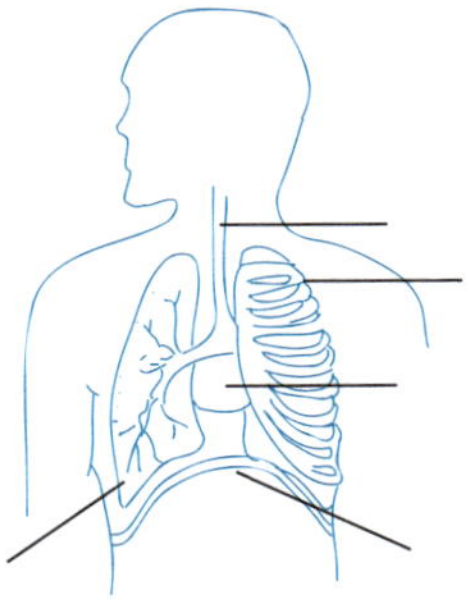

b Study the diagrams of air sacs or alveoli below then fill in the missing words in the sentence.

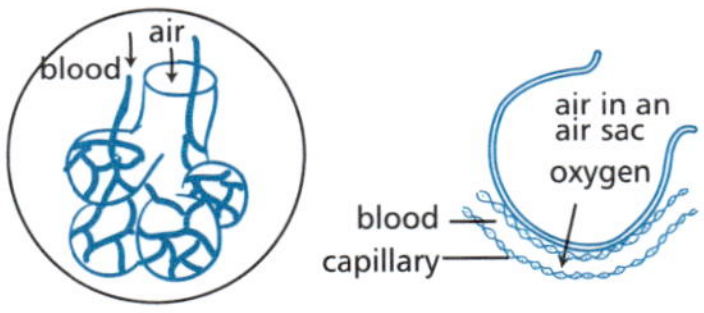

Air sacs or __________ bring air close to your blood. __________ passes from the air in the air sac onto a blood vessel called a __________. __________ __________ passes from the capillary into the __________ and is expelled from the body.

Question 6

Label the joints below with the correct description.

Descriptions:

Synovial (hinge) joint
Synovial (ball and socket) joint
Slightly moveable joint

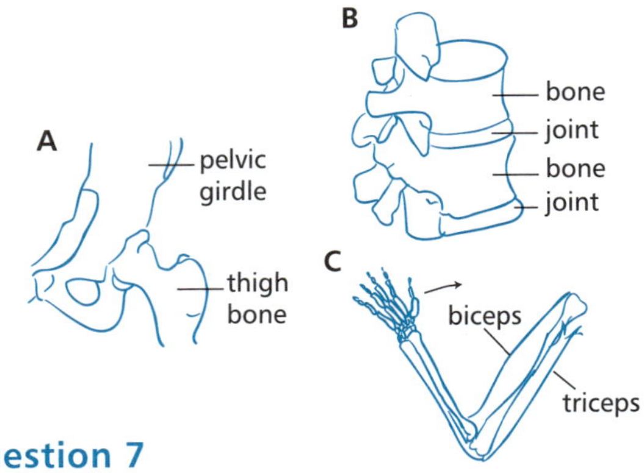

Question 7

The flow of blood around your body is called your circulation. Exercise helps your heart get stronger and your circulation to improve. Using the phrases below, complete the flow chart.

- Blood stores oxygen and gets rid of carbon dioxide.
- Your heart pumps blood to your lungs.
- Body cells swap carbon dioxide for oxygen.
- Blood leaves your lungs and travels back to the left side of your heart in veins.
- Blood from body cells returns to the right side of your heart.

Your heart pumps blood to your body cells.

Question 8

The following diagram is the fore leg of a cow. Label the bones using the terms given.

humerus, ulna, radius, scapula, digit, metacarpus, carpus

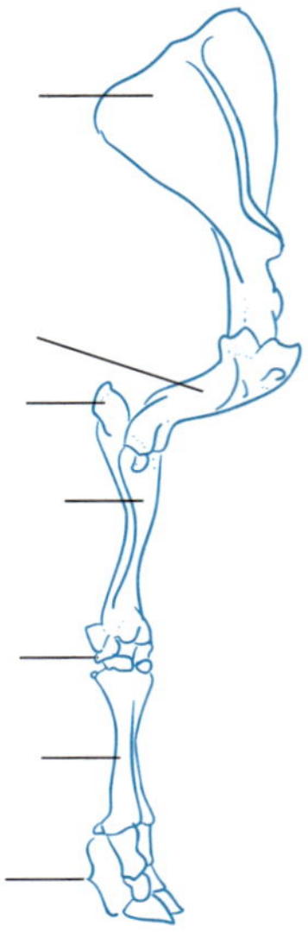

Question 9

One stage of mitosis is known as interphase.

a Name the other four stages in sequence.

____

____

b What major events take place in each of the above stages?

____

____

____

____

____

____

____

____

____

____

Question 10

Study the diagrams below. What stage of mitosis is represented by each diagram?

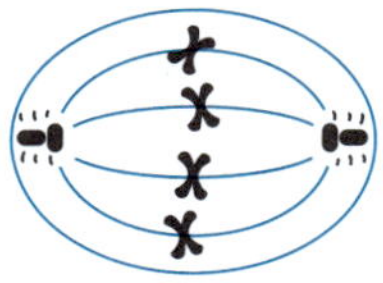

(a) ____________________

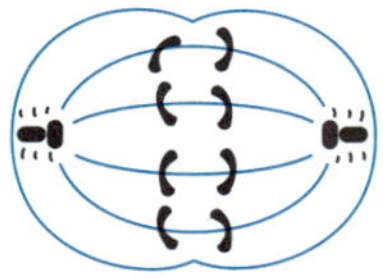

(b) ____________________

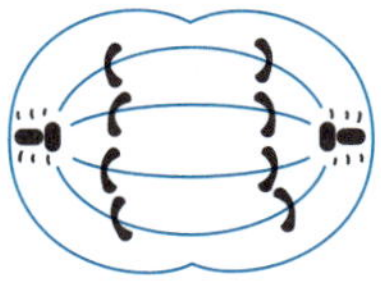

(c) ____________________

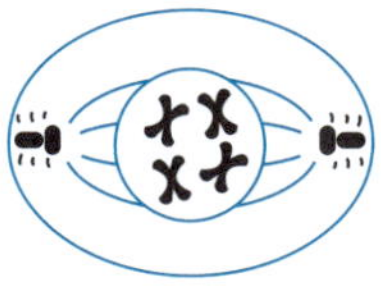

(d) ____________________

Question 11

In which of the following animal processes does mitosis occur?

a The elongation of leg and arm bones.

b The production of an ovum cell.

c The production of new skin cells.

d The increase in muscle size in athletes.

Question 12

Meiosis is a special type of ____ ________ that takes place in the ________ and ________. The normal number of chromosomes is ________. The number of chromosomes is ________ during ____________.

Question 13

a The drawing shows a closed flask containing air. Imagine you could see air in the flask. Draw how it would look in the flask.

The flask is connected to a pump and some of the air is being taken out.

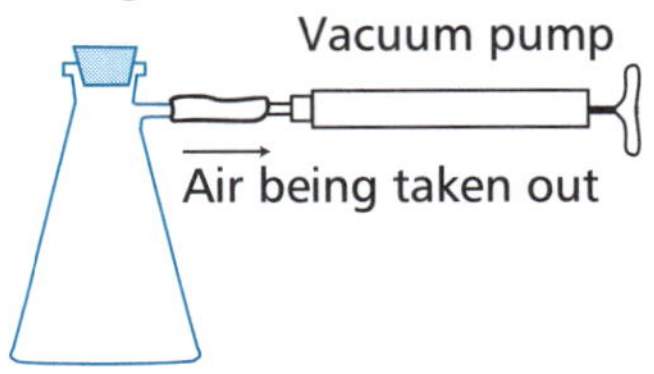

b This is the closed flask after some of the air has been taken out. Draw how the air would look in the flask now.

Question 14

a If a cup weighs 5 grams, predict the mass of B and C.

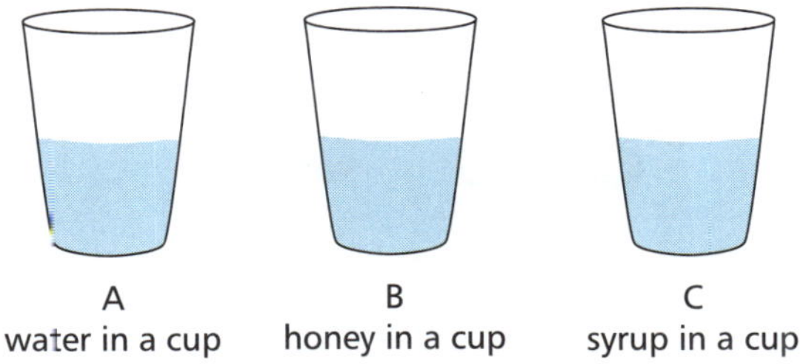

i B and C weigh the same as A.

ii B and C weigh more than A.

iii B and C weigh less than A.

b Explain your answer.

____

____

____

Question 15

Label the boxes with either:

melting

freezing

condensing

evaporating

a a whistling kettle ______

b dew forming on the grass ______

c snow on a hot day ______

d ice-cream in the freezer ______

e crayons lying in the sun ______

f a foggy street ______

g moisture on the out-side of a milk bottle ______

h puddles on the road on a sunny day ______

Question 16

Match each sentence below with the picture that it belongs to.

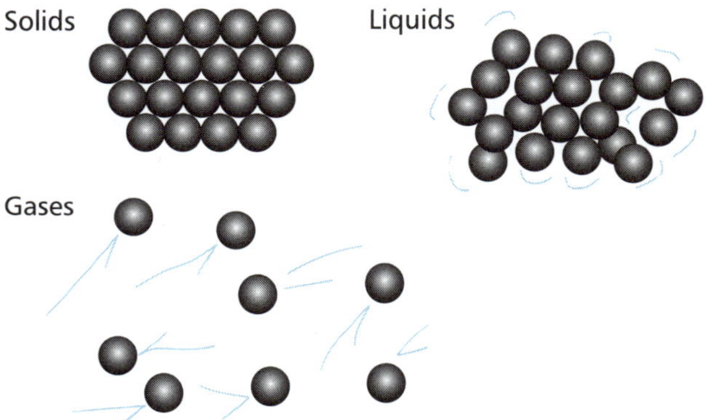

- The particles move about slowly.
- The particles are far apart.
- The particles move about very fast
- The particles are quite close together.
- The particles cannot move about but they vibrate.
- The particles are very close together.

Question 17

Which of the following statements are true?

a When solids get hot they get bigger.

b Increasing the temperature makes a solid contract.

c Solids are made up of particles packed loosely together.

d There is hardly any space between the particles of a solid.

e Particles in a solid vibrate.

f As particles get hot they get faster.

g Increasing the temperature increases the size of the vibrations.

h The larger the vibrations, the larger the space the particles take up.

i Solids contract as the vibrations get bigger.

j As vibrations decrease in size, the particles collide with each other more.

Question 18

a What has to be added to solid chocolate to change it into a liquid?

______

b What has to be added to liquid chocolate to change it back to a solid?

______

c Often flavouring is added to chocolate (eg peppermint). In which state does the chocolate have to be to add different flavours?

______

Question 19

The density of a substance is the number of particles in a set volume of the substance.

Density = Mass divided by Volume. The density of water is 1g/cm^3.

a Using this information, complete the table below.

Substance	Density (g/cm^3)	Mass (g)	Volume (cm^3)
A		4	2
B	8		4
C	2	1	
D		2	4

b Which substance has the greatest mass?

c Which has the smallest volume?

d Which two substances could be the same material?

e Which substance would float on water?

Question 20

When water evaporates it becomes a gas — water vapour. Think about drying clothing on wash day. Decide on the four factors that affect drying and explain your answer.

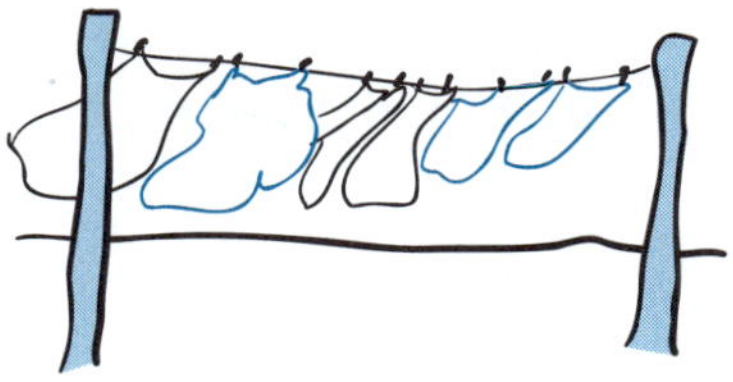

a Would you choose a dry day or a wet day?

b Would you choose a cold day or a hot day?

c Would you choose a still day or a windy day?

d Would you fold the clothes or spread them out to cover a large area?

Question 21

Some liquids pour quickly and some slowly. The viscosity of a liquid causes it to resist flow (or pouring). Liquids with a high viscosity pour slowly and are fairly thick.

A student decided to measure the viscosity of four liquids (water, honey, shampoo and oil) at room temperature. The equipment available was 4 tall jars, some marbles, a stopwatch and the 4 liquids.

Design an experiment to prove the following hypothesis: **Honey is the most viscous liquid.**

CHAPTER 5

COMPOUNDS AND CHEMICAL REACTIONS

What reaction is this?

A substance made from different elements joined together is called a **compound**. New compounds are formed by rearranging atoms rather than by creating matter. The elements are always combined in the same proportion by mass and can make millions of different substances.

Compounds can be:

- molecular compounds, or
- ionic compounds.

Molecular compounds have the following features:

- Made up of molecules.
- Usually have low melting and boiling points.
- Many are gases or liquids at room temperature.
- Most are made up of two or more non-metallic elements.
- Examples include water and carbon dioxide.

Ionic compounds have the following features:

- Made up of positive and negative ions which attract each other.
- Electrically neutral.
- Most are crystalline solids at room temperature.
- Usually formed from a metallic and nonmetallic element.
- Examples include sodium chloride and calcium carbonate.

Compounds have very different properties from the elements they are made from. For example water is hydrogen joined with oxygen. Water is a liquid and does not burn. Hydrogen and oxygen are gases that can burn or even explode.

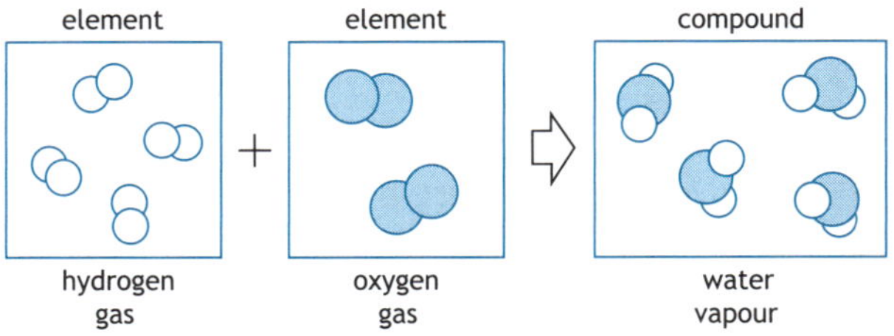

A compound consists of atoms of at least two elements

When elements join together to make molecular compounds their different atoms 'stick' to each other by sharing the electrons in their outer shells. This is called a **chemical bond**.

Naming compounds

Molecular compounds

The type and number of atoms in a compound distinguishes one substance from another. A chemical or **molecular formula** uses chemical symbols to identify the elements in a pure substance. It shows the kinds and numbers of atoms in the smallest unit or molecule of the substance. The subscript number, which follows the symbol of the element, tells how many atoms of the element are in a molecule. The number '1' is not written in a formula. For example carbon dioxide, which is made up of one oxygen atom and two carbon atoms, is written CO_2.

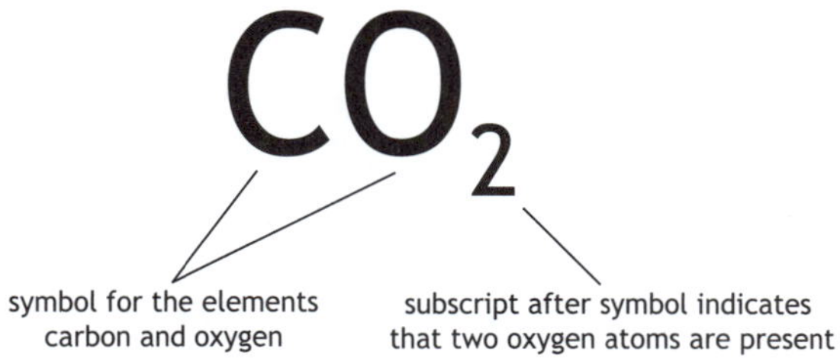

Chemical formula for carbon dioxide

Ionic compounds

The formulae of ions include not only types of atoms, but also the size and sign of the charge. The formula of an ionic compound is the formula of the cation followed by the formula of the anion. Ionic compounds are neutral.

For example:

- In sodium chloride (NaCl) there is one sodium ion for every chloride ion.
- In magnesium chloride the magnesium ion ($Mg^{2}+$) has two positive charges and the chloride ion has one negative ion (Cl–). Each $Mg^{2}+$ ion requires two Cl– ions to give a compound which is electrically neutral. Therefore the correct formula for magnesium chloride is $MgCl_2$.

You can often tell what is in a compound from its name. For example copper chloride tells us that it is a compound of copper and chlorine. Iron oxide is made up of iron (Fe) and oxygen (O) molecules.

Chemical reactions

When substances rearrange their atoms between the molecules to give different chemical substances a chemical reaction has taken place. The starting materials in a chemical reaction are known as the **reactants**, and the new substances produced by the chemical reaction are called the **products**.

For example, when we burn wood, each carbon atom in the wood combines with an oxygen molecule (two oxygen atoms) to form carbon dioxide. Because the atoms are now joined together differently, carbon dioxide has different properties from those of either oxygen or carbon.

In a chemical reaction no atoms are destroyed, nor are any created. This is called the **law of conservation of mass**.

Writing down chemical reactions

To simplify the relationship between reactants and products, and to show what happens in a chemical reaction, we write a chemical equation:

- We put what we start with on the left.
- We put what we end up with on the right.

For example:

carbon + oxygen → carbon dioxide

The arrow means to react or to make.

We can also write this equation using symbols:

$$C + O_2 \rightarrow CO_2$$

There are three types of chemical equations:

1. **Molecular equation** – shows the molecules that are taking place in the reaction.

$$HCl + NaOH \rightarrow NaCl + H_2O$$

2. **Ionic equation** – shows what is happening to each kind of ion in the reaction.

$$H^+ + Cl^- + Na^+ + OH^- \rightarrow Na^+ + Cl^- + H_2O$$

3. **Word equation** – shows what is happening in the reaction using only words.

hydrochloric acid + sodium hydroxide → sodium chloride + water

Word equations, molecular equations and ionic equations tell you what happens in a reaction. But molecular and ionic equations give much more information than a word equation.

Common types of reactions

1. Combustion

The reaction of a substance with air, at high temperatures, is called combustion or burning. Atoms of the element join up with oxygen atoms. These compounds, produced by combustion, are called **oxides**.

For example:

- If magnesium is burned, the magnesium reacts with the oxygen in the air – it gains oxygen and becomes magnesium oxide. The word equation is:

  magnesium + oxygen → magnesium oxide

  $$2Mg + O_2 \rightarrow 2MgO$$

  The ionic equation is:

  $$2Mg + O_2 \rightarrow 2Mg^{2+}O^{2-}$$

- When a fuel such as methane (CH_4) burns, it gains oxygen – it is oxidised. The word equation is:

  methane + oxygen → carbon dioxide + water

  $$CH_4 + 2O_2 \rightarrow CO_2 + 2H_2O$$

- Respiration is the release of energy from food in living cells. Glucose from food reacts with oxygen, and the carbon in glucose bonds with the oxygen. The word equation is:

  glucose + oxygen → carbon dioxide + water

  $$C_6H_{12}O_6 + 6O_2 \rightarrow 6CO_2 + 6H_2O$$

2. Corrosion

Corrosion is the reaction of a metal with another element, usually oxygen, to form a new substance. Many metals corrode or rust in moist air. The new substance, iron oxide or rust, is undesirable because it is very different from the original metal.

iron + oxygen → iron oxide (rust)

$$4Fe + 3O_2 \rightarrow 2Fe_2O_3$$

3. Precipitation

Precipitation reactions only occur in solution. They involve the mixing of two solutions and the production of a solid.

If we mix copper sulphate ($CuSO_4$) solution with sodium hydroxide (NaOH) solution they react together to form copper hydroxide and sodium sulphate. In the solution the ions are all separated from each other and moving around, so they bump into each other. When a copper (Cu^{2+}) ion bumps into a hydroxide (OH^-) ion the ions are strongly attracted and form an insoluble solid which is called a **precipitate**.

4. Neutralisation

Acids are acidic because they contain hydrogen ions (H^+). The more hydrogen ions they contain the more acidic they are. Acids producing lots of H^+ ions and are called **strong acids**, those producing only a few are called **weak acids**. An acid, therefore, is a substance that produces H^+ ions when it dissolves in water.

For example, hydrogen chloride is a colourless gas which, when dissolved in water, produces hydrochloric acid.

$$HCl \rightarrow H^+ + Cl^-$$

An **alkali** is a substance that produces OH^- ions when it dissolves in water. Like acids, alkalis can be strong or weak, depending on how many OH^- (hydroxide) ions it produces.

Acids and alkalis readily react together. If there are equal numbers of H^+ and OH^- ions then the hydrogen ions are removed by combining with the OH ions to form water and the solution is neutralised. A new substance called a **salt** is also formed.

$$HCl + NaOH \rightarrow NaCl + H_2O$$

5. Acids on metals and carbonates

- Metals and acids – Many metals react with acids, neutralising them. In this case the hydrogen ions do not combine with oxygen to form water. The hydrogen ions from the acid gain electrons from the metal atoms. Then the hydrogen atoms join in pairs to form hydrogen gas.

  zinc + sulphuric acid → zinc sulphate + hydrogen

  $$Zn + H_2SO_4 \rightarrow ZnSO_4 + H_2$$

- Carbonates and acids – A carbonate fizzes (effervesces) when it reacts with an acid. The gas produced is carbon dioxide. This reaction is often used to find out if a substance contains a carbonate.

  Sodium hydrogen carbonate, or baking powder, is used to neutralise the hydrochloric acid in our stomach if we suffer from indigestion. Farmers use neutralisation reactions to control the pH of their soil.

hydrochloric acid + sodium hydrogen carbonate → sodium chloride + water + carbon dioxide

$$HCl + NaHCO_3 \rightarrow NaCl + H_2O + CO_2$$

6. Decomposition

This occurs when a compound is broken down into its elements. Water can be decomposed, or split into its elements hydrogen and oxygen gas, by passing an electrical current through it.

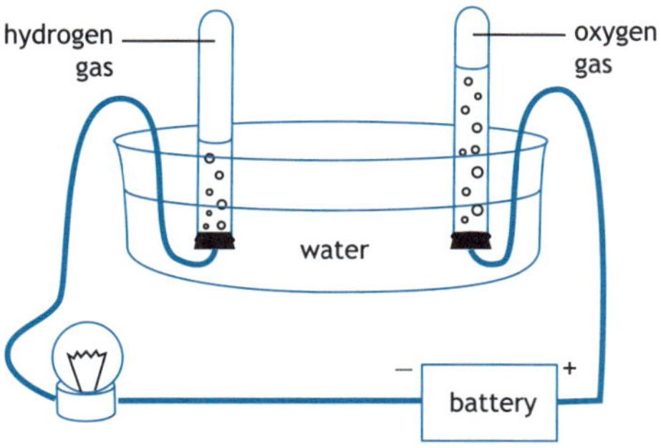

Splitting water into its elements

Glossary

Chemical bond – A link between two atoms.

Chemical reaction – A process where the chemical form of the substances is changed.

Combustion – The chemical change producing heat, in which oxygen reacts with another substance.

Compound – A pure substance made up of more than one kind of atom.

Decomposition – A process where a compound breaks down into its separate elements.

Molecular formula – A means of summarising the type and number of elements of a substance.

Neutralisation – A reaction in which an acid and an alkali react in an aqueous solution to form a salt and water.

Oxides – Compounds which contain oxygen joined to a different element.

Precipitate – The insoluble product formed in a precipitation reaction.

Precipitation – The formation of an insoluble solid when a reaction takes place in a solution.

Product – A new substance formed in a reaction.

Reactant – A substance initially present before a reaction commences.

Question 1

An element may be represented by a chemical symbol. Study the following symbol and then answer the questions.

symbol of elements

$$H_2SO_4$$

a What elements do the letters H, S and O stand for?

b What is the name of H_2SO_4?

c What do the small subscript numbers tell us?

d What does the chemical formula of a compound tell us?

e What is the smallest particle of a substance called?

Question 2

Study the following diagram, which represents the reaction between sodium and chlorine.

Write in the boxes the names of the substances you would find in each flask.

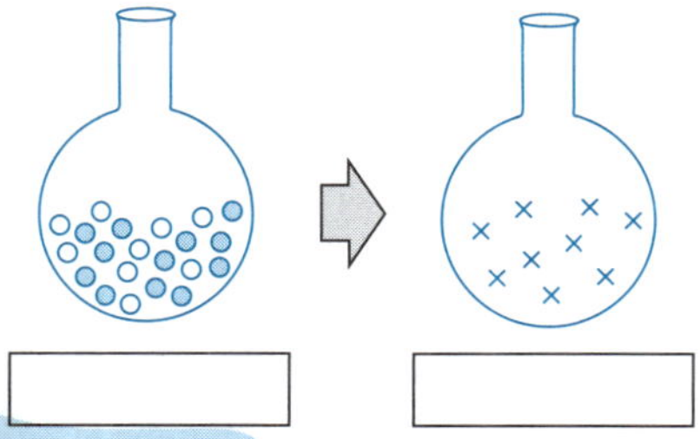

Question 3

Burning is a type of chemical reaction.

a Write down the word equation for the reaction which occurs when iron is burned.

b What two substances does the reaction begin with?

c What substance is produced?

d Write the word equation for the reaction between magnesium and oxygen.

e What is the name of the new compound formed when magnesium is burned?

Question 4

The following substances react together to form two new compounds.

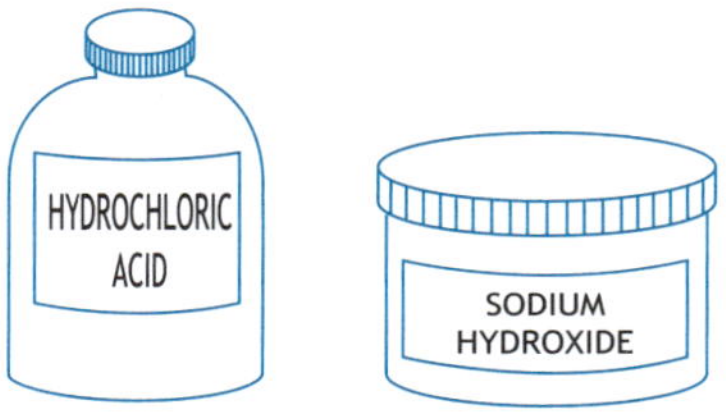

a What new compounds are formed?

b Write the word equation for this reaction.

c Write the molecular equation for this reaction.

d What name is given to this kind of chemical reaction?

e Why does it have this name?

Question 5

The diagram below shows the reaction between a metal (zinc) and an acid (hydrochloric).

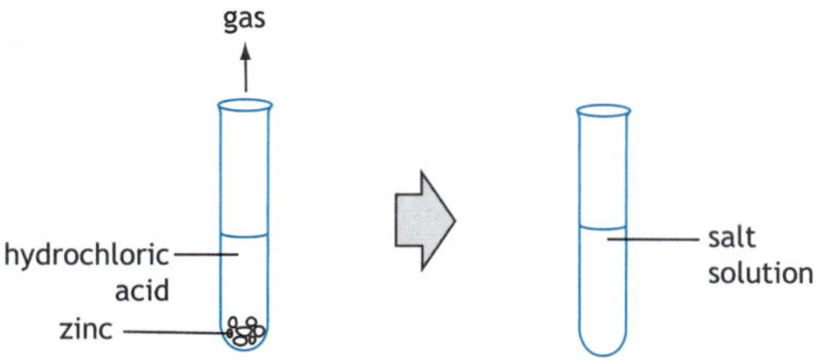

a Does the acid become neutralised in this reaction?

b What salt is formed?

c Is water formed?

d What gas is formed?

e Write the molecular equation for this reaction.

Question 6

Complete the following table.

Substance	What atoms is it made of?	Element or compound?

Question 7

The following models show two compounds:

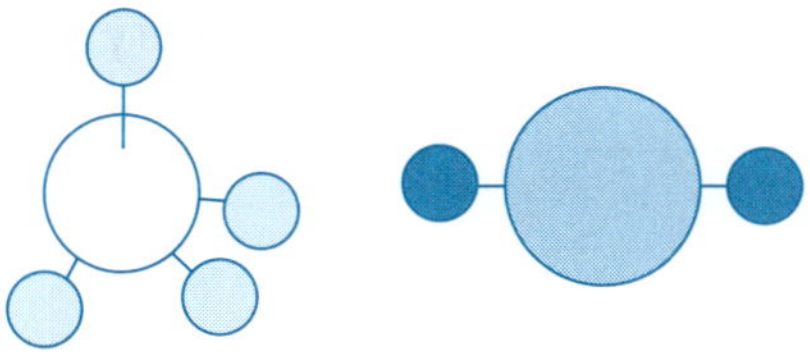

Compound AB_4 **Compound C_2D**

Which of the models below would represent a compound with the formula AD?

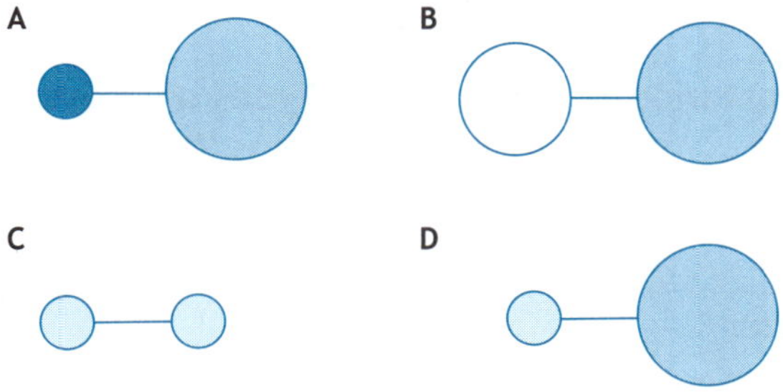

Question 8

Students carried out the following experiment to find out what causes an iron nail to rust.

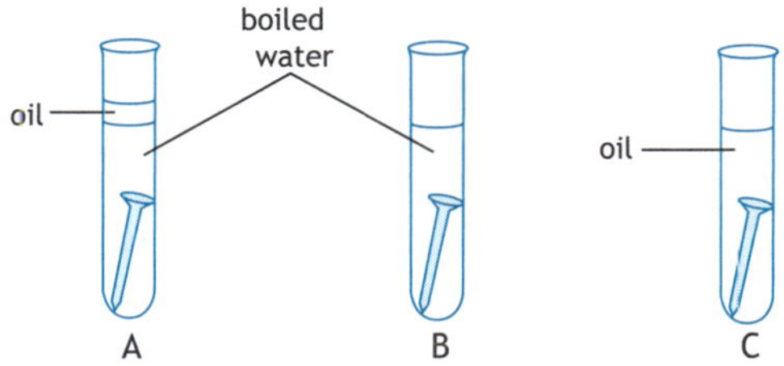

a Which nail do you think would rust?

b What is the name of the reaction that is occurring in this test tube?

c What element is the iron combining with?

d Why did they put oil on top of tube A?

e What two things does iron need to rust?

Question 9

Chemical symbols are a chemist's shorthand. Complete the table below.

Substance	Chemical formula	Types of atoms (No.)	Atoms in total
hydrogen gas	H_2	1	2
water		2	
sulphur	S_8		
methane	CH_4		
vinegar	CH_3CO_2H		
sulphuric acid	H_2SO_4		
alcohol	CH_3CH_2OH		
carbon dioxide			
nitrogen gas	N_2		

Question 10

a Write the equation for the following reaction:

Sodium carbonate powder is mixed with hydrochloric acid, resulting in the formation of carbon dioxide gas and a solution of sodium chloride.

____

b Ionic compounds are made up of __________ and __________ ions that __________ each other.

c Write the ionic formula for the following ionic compounds.

Na_2CO_3 __________

NaCl __________

Mg_3N_2 __________

K_3N __________

F_2O_3 __________

Revision questions

Question 1

The molecular formula shows the number and kinds of atoms present in a molecule of a compound.

a What kinds of atoms are in the compound H_2O (water)?

____

b How many of each atom are present?

____

Question 2

Complete the following information by filling in the missing words.

We show what happens in a chemical reaction by writing an __________. The substances we start with are written on the __________ and the ones we finish with are written on the __________. The arrow means to . Chemical reactions make __________ substances.

Question 3

The elements sodium and chlorine are both very reactive. However, the compound sodium chloride is very stable.

a What is the molecular formula for sodium chloride?

____

b Explain why this compound is very stable.

____

____

____

____

c Write the ionic formula for sodium chloride.

____

d What term is given to the sharing of electrons by atoms?

____

Question 4

Match the following terms with the correct definition.

Precipitation	The chemical change in which oxygen reacts with another substance.
Combustion	A reaction in which an acid and an alkali react in an aqueous solution to form a salt and water.
Neutralisation	The formation of an insoluble solid when a reaction takes place in a solution.

Question 5

Study the following chemical equation:

$$Na + 2H_2O \rightarrow NaOH + H_2$$

a What is the chemical formula of each product?

____

b What is the chemical formula of each reactant?

____

Question 6

Label the columns with the correct type of compound and provide an example of each.

neutral	neutral
made up of cations and anions	made up of molecules
mostly crystalline	mostly gases and liquids
made up of metallic and non-metallic elements	made up of non-metallic elements
eg:	eg:

Question 7

Complete the following information by adding the correct terms.

The magnesium ion has two positive ________ and is written ______. The chloride ion has one ________ charge and is written ______. Each magnesium ion needs ______ chloride ions to give a compound which is electrically neutral. Therefore the correct formula for magnesium chloride is ________.

Question 8

What is the law of conservation of mass?

____

____

____

Question 9

What type of chemical equations are the following?

a methane + oxygen → carbon dioxide + water

____

b $Na + 2H_2O \rightarrow NaOH + H_2$

____

c $H^+ + Cl^- + Na^+ + OH^- \rightarrow Na^+ + Cl^- + H_2O$

____

Question 10

Which of the following produces the most hydrogen ions?

a alkali

b weak acid

c salt

d strong acid

Question 11

a Water is formed when hydrogen burns in oxygen. Write a word equation for this reaction.

b Write a word equation for a combustion reaction that forms carbon dioxide.

c Write a word equation for the decomposition by electrolysis of copper chloride.

Question 12

There is a mistake in each of the following equations. Find the mistake and correct the equation.

a copper + oxygen → copper sulphide

b magnesium + oxide → magnesium oxide

c copper sulphate + sodium hydroxide
→ copper hydroxide + sodium chloride

d zinc + sulphuric acid → zinc sulphate + oxygen

e hydrochloric acid + sodium hydroxide → sodium chloride + hydrogen

Question 13

Using the correct chemical equations from the above question, name each reaction.

a

b

c

d

e

Question 14

Which of the following best describes a neutralisation reaction:

a A strong acid combining with a weak acid.

b Hydrogen ions combining with hydroxide ions.

c The reaction of a metal with oxygen.

d The splitting of a chemical bond.

Question 15

Complete the table below by adding the type of reaction involved in each example.

Example	Reaction
respiration	
rusting	
relieve indigestion	
burning fossil fuels	
managing soil pH	

Question 16

a What do we call a pure substance made up of more than one kind of atom?

b What do we call the insoluble product formed in a precipitation reaction?

c What do we call a process where the chemical form of the substances is changed?

d What do we call compounds which contain oxygen joined to a different element?

e What do we call a link between two atoms?

CHAPTER 6

THE EARTH'S STRUCTURE

What was soil before it was soil?

Planet Earth is a complex structure floating in space. It is almost round like a ball but the surface has been crumpled and broken. The centre is a fiery furnace, hot enough to melt rock.

Beneath our feet it is churning with awesome power that can make the ground tremble, form mountains, erupt volcanoes and force continents to collide or tear apart as they ride on the molten rock beneath the Earth's crust.

The Earth's layered structure

Earth's interior = core + mantle

The core has two parts:

- the inner core which is a solid iron-rich nickel ball; and
- the outer core which is a thick, elastic, molten mixture with an extremely high temperature.

The mantle is 3000 km thick and surrounds the core. It is made of very hot rock that sometimes moves around like thick syrup or molasses. This material is called **magma** and the continents float around on this thick substance.

Cross-section of the Earth

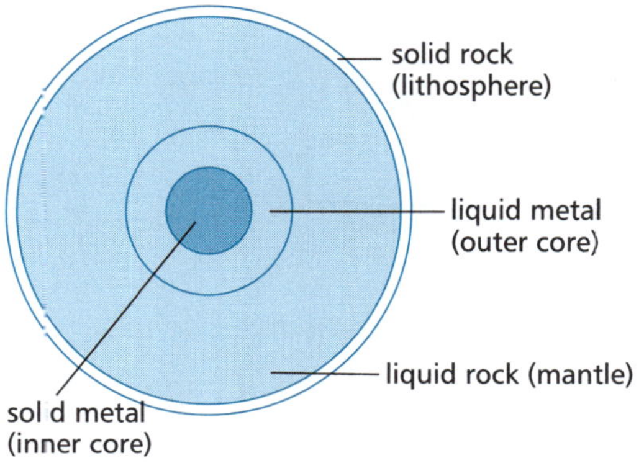

Why is the centre of the Earth so hot?

When the Earth first formed it was cold, but as the radioactive substances inside the rocks decay they release huge amounts of energy which has melted most of the rocks.

Lithosphere

The crust, or **lithosphere**, is the part of the Earth that we live on. The crust is a very thick layer (about 70 km thick) of soil and solid rock. If you have been in a cave you have been inside the crust. No-one has ever dug right through the crust.

Hydrosphere

Hydrosphere is the name for all the water on the Earth — the lakes, rivers, oceans, glaciers and ground water.

Water sometimes trickles down through cracks in the crust to the hot rocks below. If this water changes to water vapour it forces its way back to the surface. This is called a **geyser** and is a source of energy called **geothermal** energy.

What is the water cycle?

Water is always on the move. Water in lakes, rivers and oceans is heated up by the sun and **evaporates** into the air. High in the atmosphere the vapour loses this energy and it **condenses** to form tiny droplets around microscopic particles in the air. Clouds begin to form and when the droplets grow to a certain size they fall back to Earth as rain.

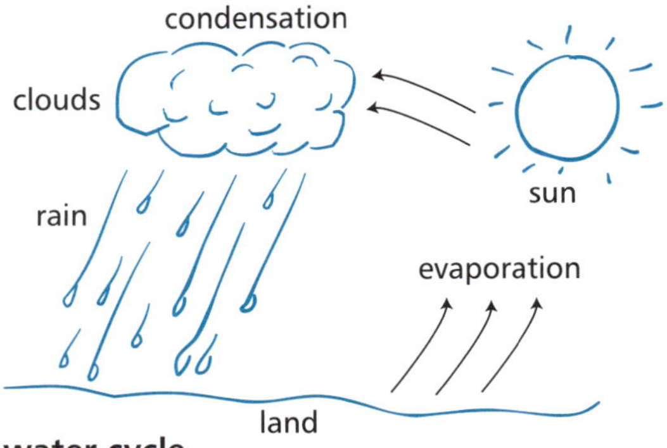

The water cycle

The moon, the sun and the hydrosphere

Ocean tides come in and go out about twice every 24 hours. Tides are caused by the gravitational pull of the sun and the moon on the hydrosphere.

When the moon and sun are pulling together on the same side of the Earth, tides are the highest. These are called **spring tides**. When the sun and moon pull against each other the tides are small and are called **neap tides**.

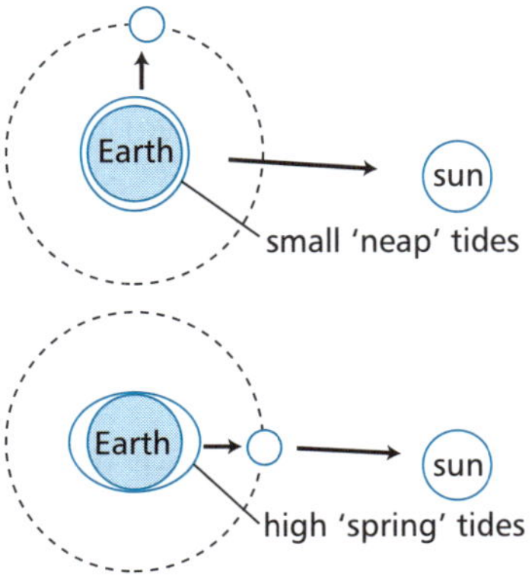

Spring and neap tides

Atmosphere

The atmosphere is a comparatively thin layer of gases which surrounds the Earth.

This gaseous layer is held in place by the Earth's gravity. This layer gives us the air we breathe and the water we drink. The atmosphere keeps us warm and protects us from the harmful rays of the sun.

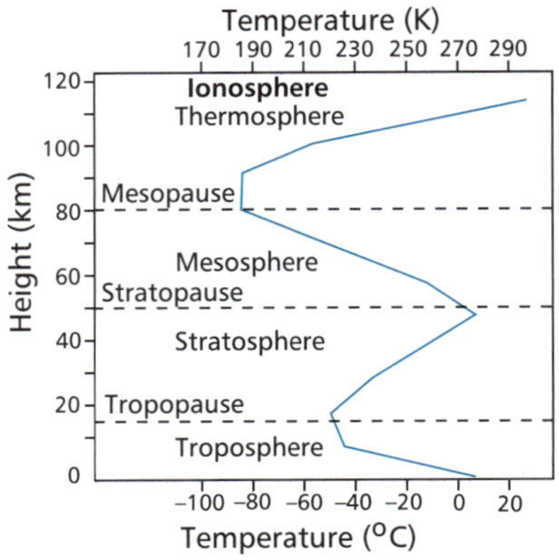

Layers of the atmosphere

The air around us is pushing constantly in every direction. This is called atmospheric pressure, or air pressure. It is caused by billions of air molecules as they zip this way and that. Air pressure is greatest close to the Earth's surface. The air is densest here because it contains the most molecules. Gravity holds molecules, particularly the heavy ones like oxygen and nitrogen, close to the Earth. Light molecules, like hydrogen and helium, move further away from the Earth as gravity does not pull on them as much. As we move away from the Earth's surface there are less and less molecules until there are none at all.

The lowest layer of the atmosphere is called the **troposphere**. The sun warms the Earth below it and the heat causes the troposphere to churn around giving us our weather — wind, rain, cyclones, clouds and snow.

What is in the air?

Nitrogen makes up 78% of the atmosphere and oxygen is 21%. The other 1% is made up of carbon dioxide, water vapour and small amounts of helium, hydrogen, neon and ozone.

What is ozone?

Ozone is a colourless gas whose particles are made up of three oxygen atoms joined together. Ozone forms the layer that keeps out most of the sun's damaging ultraviolet rays.

Gases like CFCs (chlorofluorocarbons) from refrigeration, manufacturing and cleaning are 'eating' holes in the ozone layer. Less ozone will cause more sunburn and more skin cancer in animals on Earth.

What is the greenhouse effect?

'Global warming' or the 'greenhouse effect' is the gradual warming of the Earth because carbon dioxide and other gases stop the heat escaping back into space.

The atmosphere is supposed to keep some heat in close to the Earth. The problem is that over the last 100 years the burning of fossil fuels has doubled the level of carbon dioxide in the atmosphere. Humans have also cut down millions of trees that could have helped to absorb the carbon dioxide.

Scientists now believe that the increasing global temperature could cause major changes in weather patterns and in sea levels.

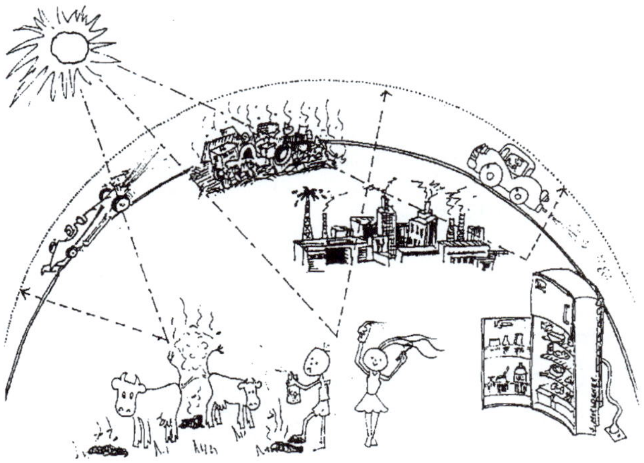

The greenhouse effect

Rocks

Different types of rocks

The crust of the Earth is made up of many different types of rocks, made in different ways. Rocks are mixtures of different minerals. Rocks are classified into three groups.

- **Igneous rocks** were once so hot that they were like a thick liquid in the mantle of the Earth. This molten rock, called magma, forced its way to the surface of the Earth where it cooled and solidified. The longer rocks take to solidify the larger the crystals are. Basalt and granite are examples of igneous rock.

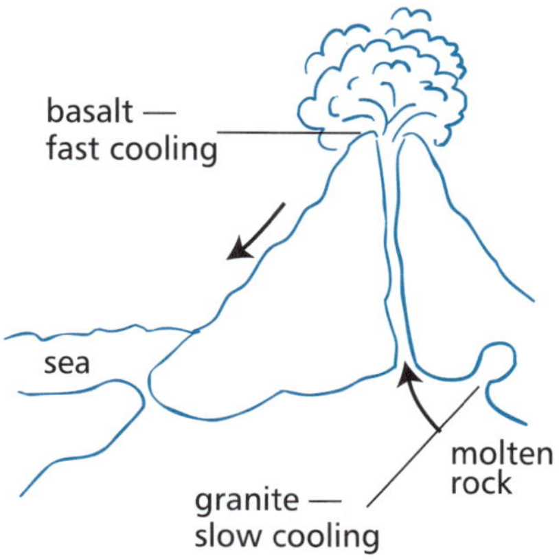

Formation of igneous rock

- **Sedimentary rocks** begin as bits of rock, mud, sand, shell and bone falling to the bottom of the sea or a lake. Examples are limestone, shale and sandstone.
- **Metamorphic rocks** have been heated and put under pressure. Over thousands of years, these rocks slowly change deep in the Earth's crust. Slate and marble are metamorphic rocks formed from shale and limestone respectively.

Breaking down the Earth's crust

The Earth's weather breaks up rocks — this process is called **weathering**. Weathering shapes the Earth's crust and produces the soils that we use to grow our food. To produce new soil from rocks takes thousands of years.

Weathering mechanisms

- Heating rocks causes them to expand and cooling causes them to contract. Constant heating and cooling can make rocks crack.
- Rainwater expands when it freezes, and if it gets into the cracks of rocks it can crack the rock further, or even break pieces off.

Weathering of rock by water

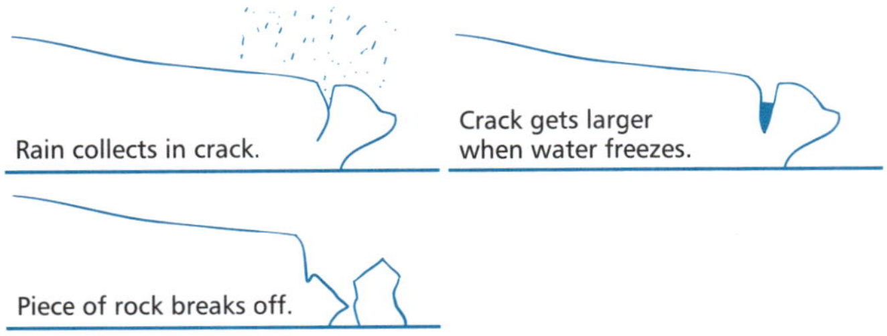

- Bits of dust and sand blown by the wind can wear away rocks.
- If seeds germinate in the cracks of rocks their roots force the cracks open further.
- Acid gases in the air also attack rocks. Carbon dioxide dissolves in water to make a weak acid which reacts with some rocks like limestone, chalk and marble. Over thousands of years this reaction can dissolve the rocks away.

Moving soils

Once the rocks have been weathered into particles or soil, natural forces move them about. This process is called **erosion**.

- Windstorms can swirl away huge amounts of dry soil.
- Glaciers scrape rocks along the ground for long distances.
- Running water causes creek banks to erode away, gullies to form and hillsides to slip.

Over millions of years erosion has changed the shape of the Earth. Erosion levels mountains and helps form sedimentary rocks.

Looking after the environment

We live on the Earth's crust, but this is only part of our environment. The Earth's water and its atmosphere are also part of our environment. How we live and what we do affects all these parts.

We are beginning to realise that acid rain, polluted air and water, and accelerated erosion of soils is damaging the way we live. We must look after all parts of our environment.

Glossary

Crust
Earth's solid, outermost layer (up to 70 km thick).

Condensation
The turning of invisible water vapour into water drops and ice.

Core
The incredibly hot metallic centre of the Earth which is thought to be solid on the inside and liquid on the outside.

Erosion
Movement of rock fragments by glaciers, wind, running water and waves.

Evaporation
The loss of water from the Earth's surface as water vapour.

Geothermal
Heat energy from the centre of the Earth.

Geyser
A mixture of water vapour and boiling water forced to the surface through cracks in the Earth's crust.

Hydrosphere
All the water on the planet — oceans, seas, rivers, lake etc.

Igneous rock
One of the three main rock types, forming when magma cools.

Lithosphere
The Earth's shell which includes the crust and the rigid upper part of the mantle.

Magma
Hot molten rock under the Earth's surface.

Mantle
The layer of the Earth's interior between the crust and the core.

Metamorphic rock
Rock made by the reforming of other rocks under intense heat and/or pressure.

Glossary (continued)

Neap tides
Small tides occurring when the moon and the sun are at right angles to the Earth.

Sedimentary rock
One of the three major rock types, made from sediments laid down in beds, mainly on the sea floor.

Spring tides
Very large tides formed when the Earth, the moon and the sun are in a straight line.

Weathering
The breakdown of rock when it is exposed to the sun, the wind and the rain.

Science skills

Question 1

A mine supervisor checks the temperature of his mine every 100 metres to the bottom level of the mine.

He records his data in the table below:

Mine depth (metres)	Mine temperature (°C)
0	20
100	22
200	24
300	26
400	28
500	30
600	32
700	34
800	36
900	38
1000	40

a Draw a graph and plot the data.

b What is the temperature at ground level?

c What is the temperature at the bottom level?

d How deep is the mine?

e Why does the temperature increase as the supervisor goes deeper?

Question 2

Match the following labels with the numbers 1–5 on the diagram below.

sedimentary rock, basalt, granite, metamorphic rock, magma

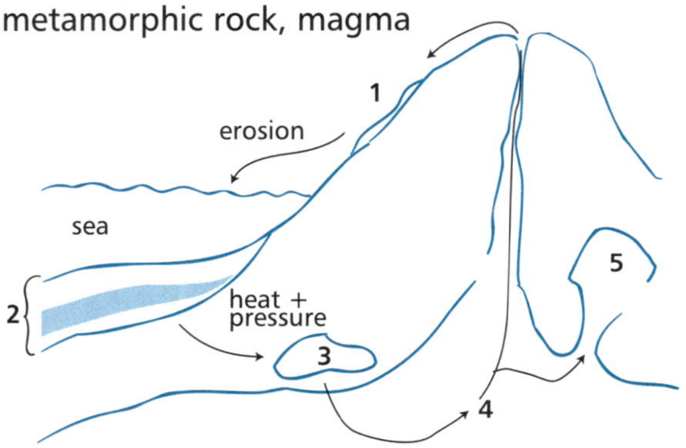

Question 3

Study the diagram below and write the correct words in the blank boxes to describe how sedimentary rock is formed.

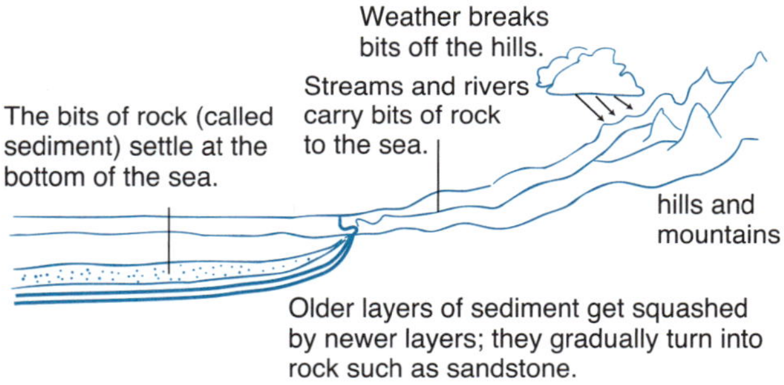

Bits of rock are broken off hills and mountains by the weather.
a
b
c
Over millions of years the bottom layers of sediment stick together to make sedimentary rock.

Question 4

a Look at the rocks below and sort them into the three groups that scientists sort rocks into.

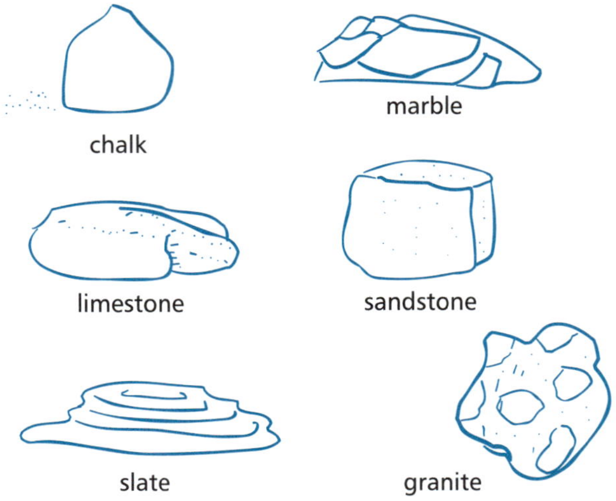

b Which three rocks are made of the same material?

c Which rock are many old buildings made of?

d Which rock can be split and used for floors in homes?

Question 5

Some students decided to investigate the effects of rapid temperature change on rocks. They set up the experiment below. What changes would they expect to see in the rock?

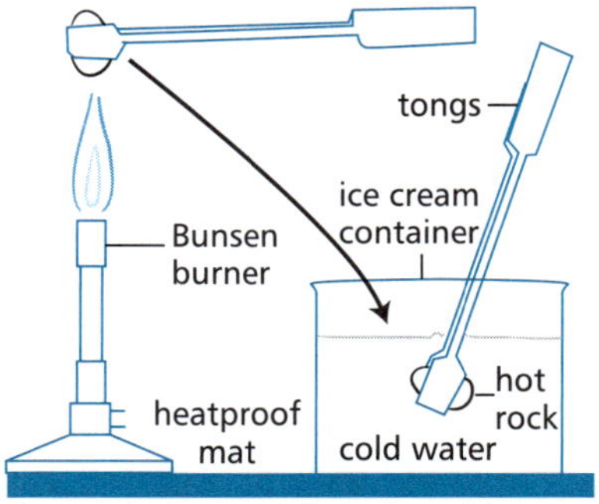

Plunge the hot rock into the cold water

Question 6

Sedimentary rocks often contain calcium carbonate which reacts with acid to form bubbles of carbon dioxide.

Which of the following rocks will react with the vinegar (ethanoic acid)?

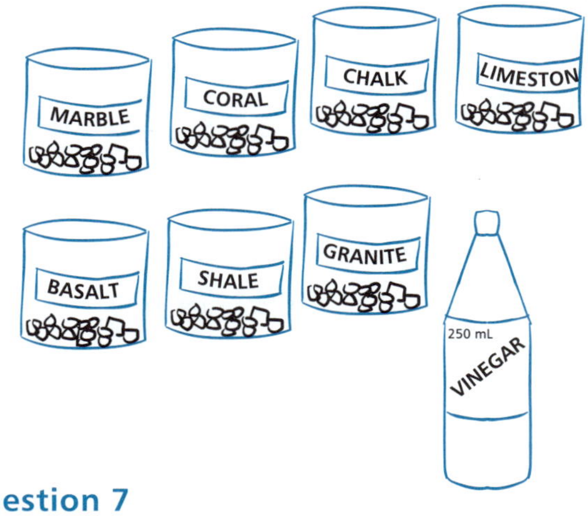

Question 7

Students decided to demonstrate the greenhouse effect by setting up the following experiment.

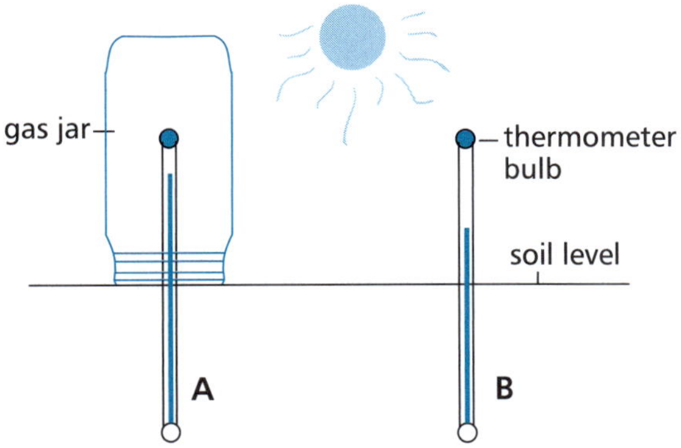

a Which thermometer will record the highest temperature?

b What does the glass jar represent?

Question 8

Complete the flow chart below by adding the following types of weathering to the boxes.

Physical weathering — wind, water, ice

Chemical weathering — acid, rusting

Biological weathering — plants, animals

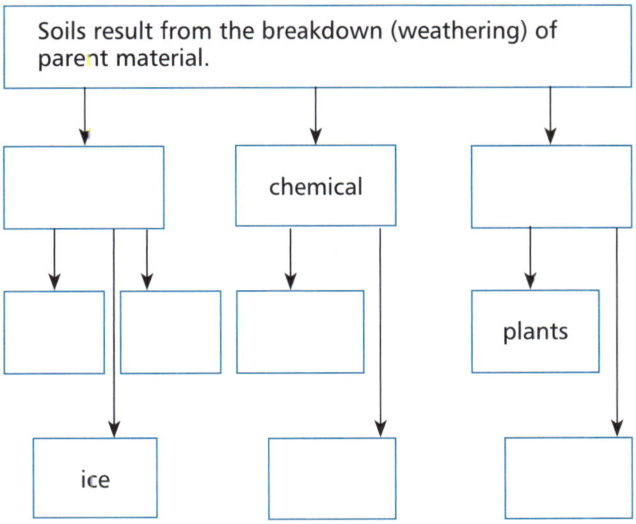

Question 9

Label the following diagram of a cross-section of the Earth.

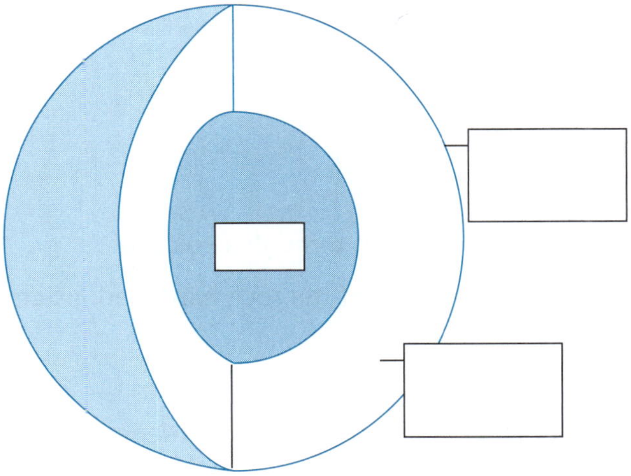

Revision questions

Question 1

Just below the crust of the Earth is the —

a inner core

b mantle

c outer core

d atmosphere

Question 2

The temperature is highest in the Earth's —

a crust

b mantle

c outer core

d inner core

Question 3

Look at the diagram below.

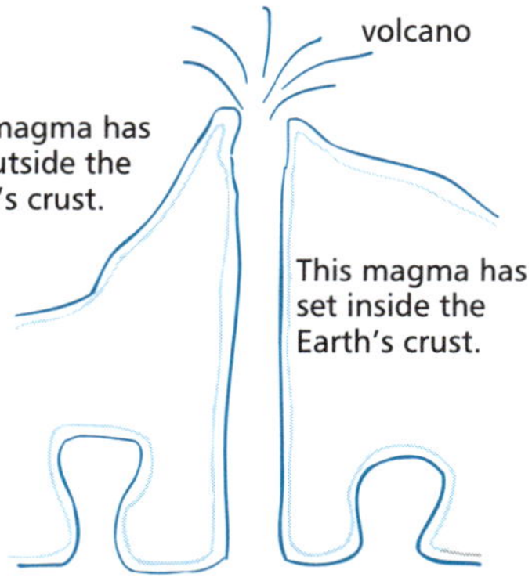

Cross-section through a volcano

a What is the name of the rock hardened inside the Earth's crust?

____

b What is the name of the rock hardened outside the Earth's crust?

____

c What is the name given to rocks such as basalt and granite?

____

Question 4

Some rocks are changed by heat and pressure in the Earth's crust.

a What does heat change limestone to?

____

b What does pressure change shale to?

____

c What name is used to describe these types of rocks?

____

Question 5

Fill in the correct words in the sentences below.

When rocks are worn away by the weather, we call it ______________. Changes in temperature from hot to cold can make __________ in rocks. If water gets into these cracks and freezes, it makes them bigger because water __________ when it freezes. Sometimes frozen water makes bits of rock __________ off.

Question 6

a How is acid formed in the atmosphere?

____

____

b What happens to limestone and chalk when this acid falls on it?

____

____

c This is called _____________ weathering.

Question 7

If you wanted to find out if an igneous rock had been formed by slow or rapid cooling you would look for:

a how hard the rock is

b the size of its crystals

c the colour of the rock

d the size of the rock

Question 8

a What are the two main gases in the atmosphere?

b Which gas makes up most of the air?

c Which gas combines with water to form carbonic acid?

Question 9

a List the four things that the atmosphere does that allow life to exist on Earth?

b List three ways that human health is at risk from a thinning of the ozone layer.

Question 10

a What is the atmosphere?

b What is the name of the layer of the atmosphere in which we live?

c What causes the greenhouse effect?

d Label all the oxygen atoms in the diagram below to show how ozone is formed.

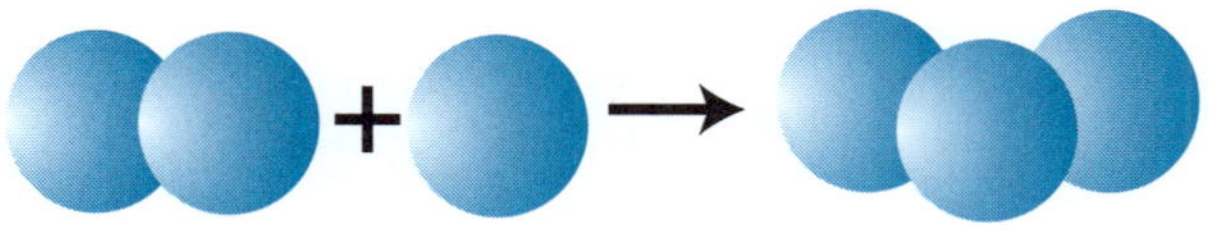

A B

e Which group of oxygen atoms, A or B, is an ozone molecule?

Question 11

Fill in the boxes with the words and phrases below to complete the water cycle.

Evaporation	Rain
River	Ocean
Condensation	Glacier
Lake	Ocean

Question 12

If you freeze a full bottle of water, the bottle will crack. How is this related to the formation of soils?

Question 13

Study the diagram below and answer the following questions.

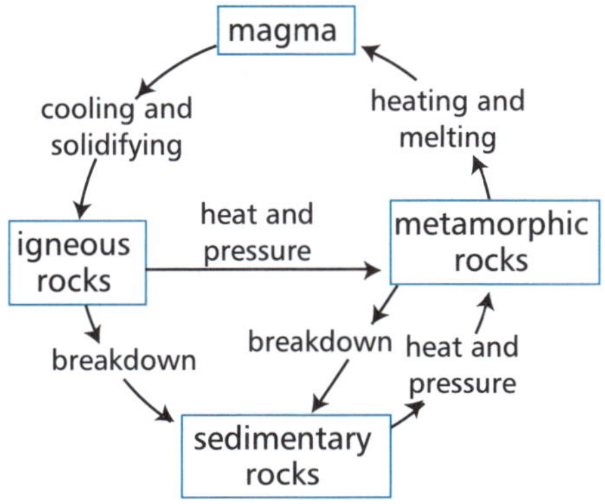

a Which two rock types were originally igneous rocks?

b Which two rock types form sedimentary rocks?

c Which rock type results from the cooling and solidifying of magma?

d What causes sedimentary rocks to become metamorphic rocks?

e Can metamorphic rocks become sedimentary rocks?

Question 14

The diagram below shows the layers in the atmosphere.

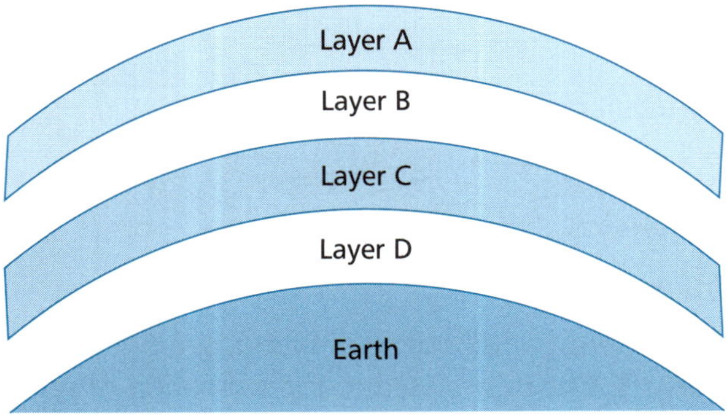

The thermosphere is the layer closest to the sun so it is the hottest.

The mesosphere is below the thermosphere.

The stratosphere is above the troposphere.

The troposphere is the layer where we find life.

Which choice places the layers in the correct order from the Earth's surface?

a thermosphere, stratosphere, troposphere, mesosphere

b stratosphere, mesosphere, thermosphere, troposphere

c troposphere, stratosphere, mesosphere, thermosphere

d troposphere, stratosphere, mesosphere, thermosphere

Question 15

Complete the following sentences with the correct words.

One of the greenhouse gases is __________ __________. Scientists think that the amount of carbon dioxide in the air increases because we burn __________ __________. An increase in the greenhouse effect will raise the Earth's average ______________.

Question 16

Write down one sentence to say what makes up each of the following parts of the Earth.

a the atmosphere

b the crust

c the magma

d the core

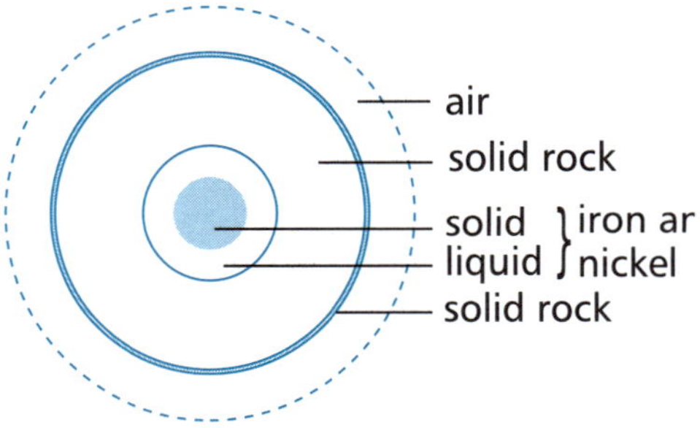

Question 17

The following sentences are about igneous, metamorphic and sedimentary rocks. Are they true or false?

a Sedimentary rocks are made from bits of other rocks that have melted together. ______

b Sedimentary rocks are made from grains which have been pressed together. ______

c Igneous rocks can be formed in volcanoes. ______

d Marble is a metamorphic rock. ______

e Sandstone is an igneous rock. ______

Question 18

Will the following arrangement of sun, moon and Earth produce spring tides or neap tides?

____

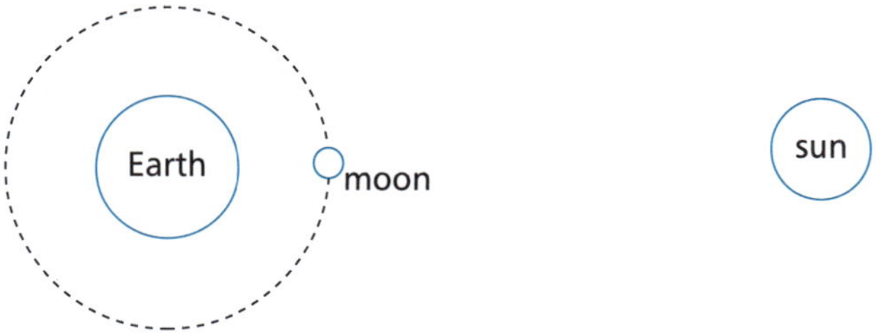

Question 19

Complete the following sentences with the correct words.

The air around us is pushing constantly in every direction. This is called _____ _________.

This pressure is greatest at _____ _________ because the air is densest at this level.

Question 20

a What part of our planet is responsible for the Earth's weather patterns?

____

b What is weathering?

____

____

____

c What is erosion?

____

____

____

d What process is responsible for the breaking up and wearing away of the Earth's rocks?

____

CHAPTER 7

ENERGY (SECTION 1)

If you scream in space can anyone hear you?

Matter moves

- Matter is moving all around you.
- It takes energy to make matter move.
- Nothing can happen without energy — to move things (a rocket to fly); to change things (butter to melt).
- Energy makes things happen.
- It is energy that creates all forces.

What then is energy?

Nobody really knows except that it is an invisible force.

Where does energy come from?

All the energy on the Earth comes from the sun. It is stored in the Earth's matter. Scientists talk about two main types of energy:

- **Kinetic energy** — This is the energy something has because it is moving (push bike, golf ball, strong winds). Remember the kinetic theory in chapter 2 (p. 15)?
- **Potential energy** — This is stored energy. It is something waiting to happen (coal, wood, food, mouse trap, wound-up rubber band).

What forms of energy are there?

There are many forms of energy and they can be classified into the two groups above.

Energy in action (kinetic)

- **Light energy** — Light energy from the sun makes plants grow, and burns your skin. It can be used to make photographs or to cook food in a solar barbecue.
- **Sound energy** — Sound energy is a special kind of kinetic energy. It is the energy of the air as it vibrates. Small amounts of sound energy can cause us to hear sounds and large amounts can break a window.
- **Heat energy** — Heat energy can make things happen. The heat from burning petrol can make a car go; the heat from a boiling kettle can make a whistle blow.
- **Electrical energy** — This energy pushes an electric current through a conductor. It makes lights work, runs computers, fans and toasters.

Stored energy (potential)

- **Chemical energy** — This is energy stored in chemicals and is often released when they are burned (fuels, food, wood).
- **Nuclear energy** — This energy can be stored inside atoms. It can be released from some atoms (uranium) in nuclear power stations. Nuclear energy stored in hydrogen atoms is where the sun's energy comes from.
- **Elastic energy** — This energy is stored in matter that can be stretched, squeezed or bent. A stretched trampoline, a compressed gas or a bent pipe can store energy until the force holding it is released.

- **Gravitational energy** — This is stored when an object is lifted above the surface of the Earth. If it falls back to Earth then this potential energy is changed. A roller coaster, a falling ball and a person diving from a diving board are all examples of gravitational energy.

More about kinetic energy

You need light to see things.

Light from the sun travels in straight lines and if they hit something they pass right through (glass), bounce off (mirror) or are absorbed (bitumen).

If light rays pass right through an object you cannot see it. If some, or all of the light bounces off an object you can see it. This is called reflection. Light bounces off in all directions if the object is not smooth.

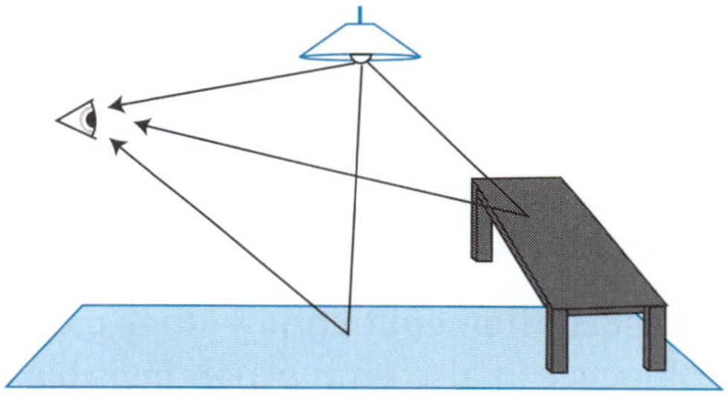

Light bulb sends out light which is reflected into your eyes.

You can't see things very well when it is dark, because no light rays are bouncing off.

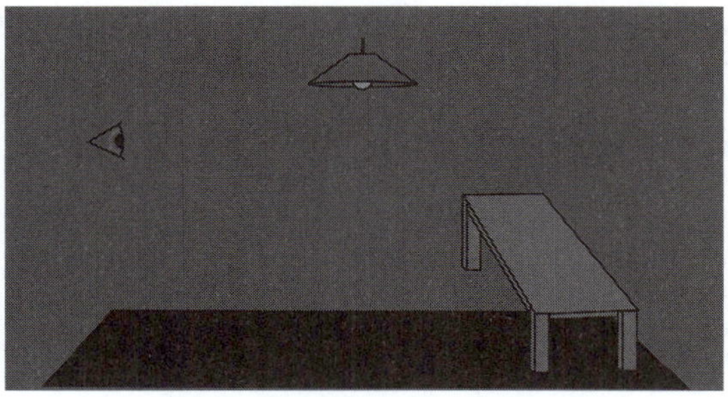

No light, no reflection, cannot see things

You can't see around corners either because no light rays are bouncing off.

Light rays cannot go through most solid things, so we usually get a shadow. On a cloudy day however the clouds let some light through and bounce some back.

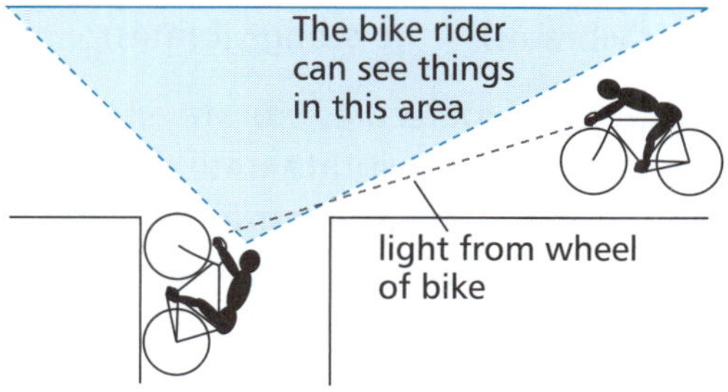

Light travels in straight lines.

Shiny surfaces like mirrors bounce (reflect) light back off at the same angle that it hits the mirror so we can see our reflection.

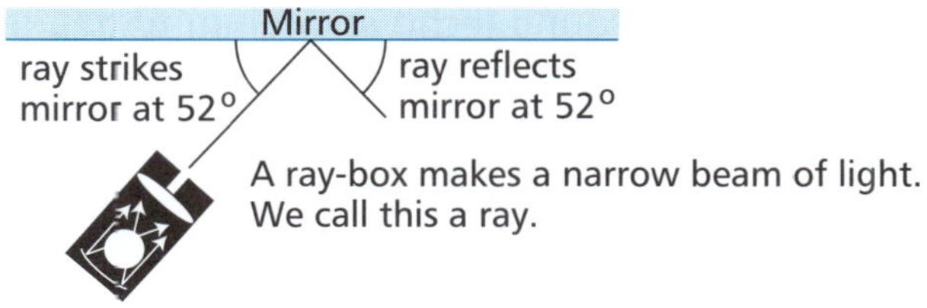

What colour is light?

The light from the sun is white. White light is not just one colour, but lots of colours all mixed up together. You can see these colours when raindrops bend the light and split it up into all the colours of the rainbow.

Why do things look coloured?

Different objects absorb and bounce back different amounts of each colour from the sun's light. An object will appear whatever colour it does not absorb. For example, a post box looks red because it bounces back mostly red light and absorbs the rest.

Some things are not coloured, they are black (absorb all light) or white (bounce all light back).

How do we hear things?

We need sounds to hear. Sounds are made when things vibrate. Different sounds are made because different types of matter vibrate at different rates.

Saxophone
(air vibrates)

Guitar
(string vibrates)

Energy makes air molecules vibrate and they make more molecules vibrate until the vibrating air molecules travel to your ear and collide with your ear drum. The ear drum vibrates, passing the **vibration** onto small bones in the inner ear. The nerves in your ears tell your brain you are hearing something.

The loud speaker in a radio makes the molecules of air vibrate creating a sound.

Sound can also be reflected. Have you heard an echo? Ships use sound (echo-sounding) to gauge the depth of the sea.

Sound can also travel through solids and liquids. Sound cannot travel through spaces (vacuums).

What do 'hot' and 'cold' mean?

When matter is heated its atoms **absorb** energy. Its molecules vibrate faster and it becomes hotter. The particles of a cold object have little energy.

Thermometers are used to measure **temperature**.

Transferring heat

Heat is energy that can be transferred. Heat will always transfer from a hot to a cold object. A substance can give off heat or absorb it. The amount it gains or loses depends on:

- the kind of substance
- the amount of the substance
- the temperature change that takes place.

There are three types of heat transfer.

1 **Conduction** is the process whereby heat is transferred in a solid. The atoms closest to the heat source absorb the energy first. As heat flows along an object, the energy of the vibrating atoms is passed on to nearby atoms, causing them to vibrate, until the whole substance is hot. Some materials, called conductors, pass on heat quickly (metals) and some materials, called insulators, pass heat on slowly (wood, glass).

2 **Convection** is the movement of heat through a liquid or through a gas. Heat is energy, not matter, so it cannot move anywhere. What actually happens is that the heat energy makes some of the particles move faster than the rest. These faster particles move through the liquid or gas, taking the heat to other places in the liquid or gas. This process cannot occur in solids because atoms in solids cannot move around.

Convection and conduction are processes of heat transfer.

3 **Radiation**
This method of heat transfer does not need particles. That is why radiation can pass through empty space. Radiation is a special type of energy called electromagnetic energy. Light is one example of electromagnetic energy. Microwaves, infra-red waves, radio waves and television waves are other examples.

Hot objects often emit light — they glow red. Hot objects can also emit other 'invisible' electromagnetic waves or radiation. The radiation travels through space until it hits some particles. The energy is then absorbed, making the particles move faster as they get hotter. You feel hot lying on the beach in the sun because the radiation from the sun makes the molecules in your skin move faster.

Controlling heat transfer

In our daily lives there are many examples of things that prevent heat gain or loss. Wetsuits and thermos flasks prevent heat loss, while white coloured cars reflect the heat. Such objects prevent heat transfer, so hot things stay hot and cold things stay cold.

What is electrical energy?

We notice the effects of electrical energy when we turn on a power point to start the TV, a vacuum cleaner, or any other electrical appliance. Electricity flows through wires to where it is needed and enters the appliance when it is switched it on.

Electrical current is like a stream of electrons (sub-atomic particles) flowing through the wires to the appliance.

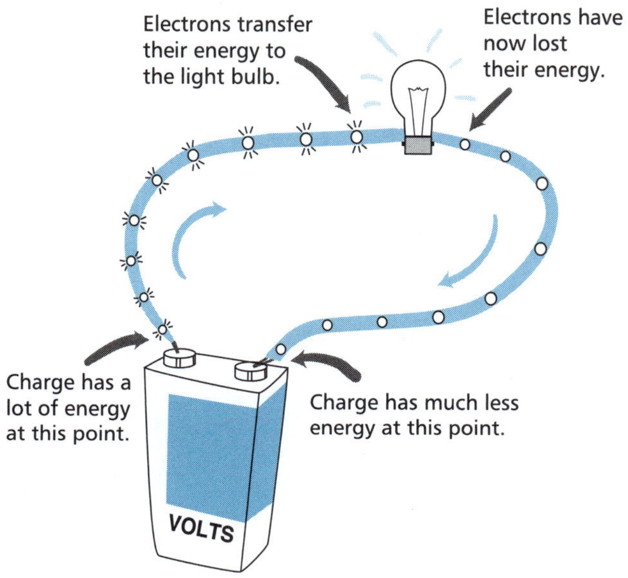

An electrical circuit

An electrical circuit must be complete before a current will flow. If there are no gaps and the switch is closed the energy will flow.

Materials that carry electric current well are called **conductors**, those that don't are called **insulators**. The wire in a circuit is usually copper as it is a very good conductor and carries the sub-atomic particles, called electrons, around the circuit quickly.

More about potential energy

If an object is to move away from the Earth it needs energy. If the object is then dropped, this saved energy is released in the form of motion (kinetic energy). The energy stored in an object depends on:

- how high the object is raised
- the mass of the object.

For example, a heavy ball will roll down a slope faster than a light ball.

Chemical energy is stored in food and other fuels. Food is burned inside our bodies to produce the energy we need for the day's activities. Chemical energy is released in chemical reactions.

Elastic energy is stored in a stretched elastic band like in a bow and arrow. When the energy is released the band will fly just as the arrow flies from the bow.

Nuclear energy is stored in the centre of an atom. It is released from uranium in nuclear power plants. Many countries use nuclear energy to produce electricity.

Energy changes

Many of the things we do depend on energy changing from one form to another. A wood fire produces light energy and heat energy. An electric drill, and hair driers produce kinetic, sound and heat energy. Only about 23% of the stored energy in petrol goes to moving the car — the rest is used to produce sound and heat energy.

During any energy changes some energy is always lost as heat energy.

The principle of conservation of energy

Energy can be changed from one form to another but it cannot be created or destroyed.

Glossary

Absorb
When something 'soaks up' energy (light) instead of reflecting it or letting it pass through.

Chemical energy
The energy stored in chemicals (food, fuel).

Conductor
A substance that energy can pass through.

Convection
This is the process that occurs when heat is transferred in liquids and gases by moving particles.

Energy
Energy makes things happen or causes things to happen.

Insulator
A substance that will not let energy pass through it.

Kinetic energy
The energy that a moving object has.

Potential energy
Energy stored in an object because of its position, or because of its condition.

Radiation
The process of heat transfer that does not need particles.

Reflection
Light or sound bouncing off whatever it strikes.

Temperature
How hot or cold something is. Measured with a thermometer.

Turbine
A motor in which a wheel is turned around when kinetic energy is transferred to it by moving air, water or steam.

Vibrations
To-and-fro movements that produce sounds.

Question 1

Steam is sometimes used to drive the turbines that produce electricity.

Write beside each picture the type of energy that the parts of a model turbine have.

a C_______ energy
in the candle wax

b H _______energy
in the candle flame

c K_______ energy
of a jet of steam

d K _______ energy
of spinning turbine

Question 2

Draw up a chart and list 5 good conductors and 5 poor conductors.

Question 3

Some students were investigating the way heat is conducted along a metal rod. They set up the equipment below and recorded how long it took for each drawing pin to fall off.

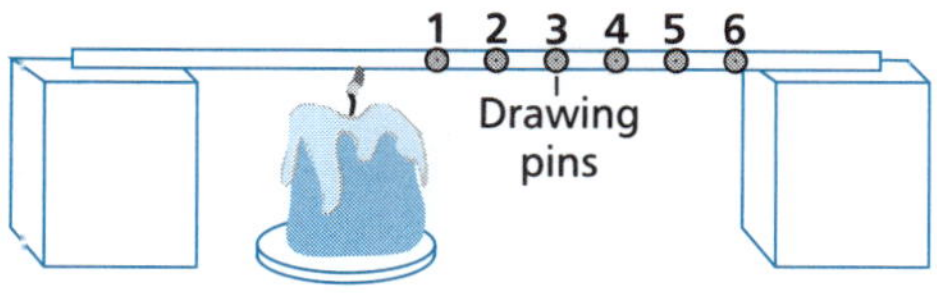

Data

Pin number	Time taken (minutes)
1	0
2	1
3	2
4	3
5	4
6	5

Draw a graph from the results and write a description of what the graph tells us. Is the graph line straight or curved?

Question 4

Look at the diagram below, and draw lines to show the two ways bike rider A can see bike rider B.

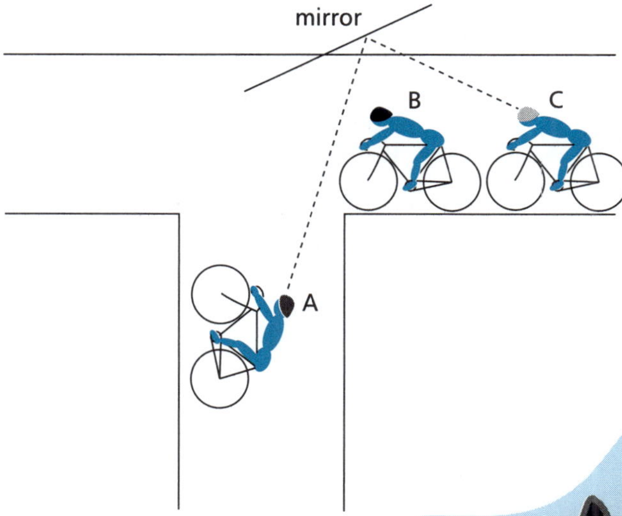

Question 5

A periscope can be made using two mirrors. The diagram shows this.

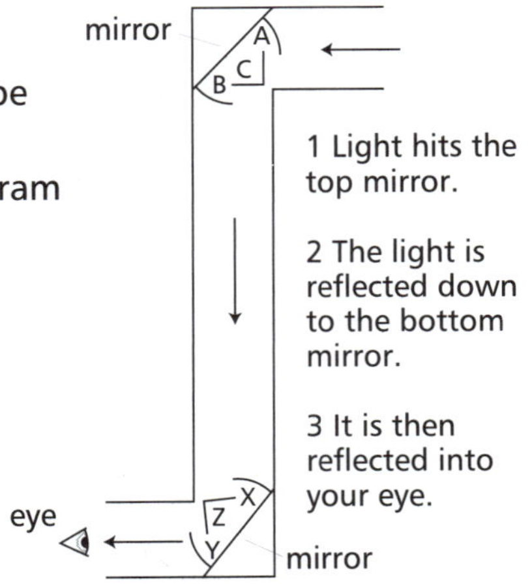

a Complete the table below.

Angle	Degrees
A	45
B	
C	90
X	
Y	
Z	

b At each mirror the light changes direction by an angle of _____ degrees. It strikes the mirror at _____ degrees and is reflected at _____ degrees.

Question 6

a Look at the diagram below. Draw the shape of the shadow on the screen and shade it in.

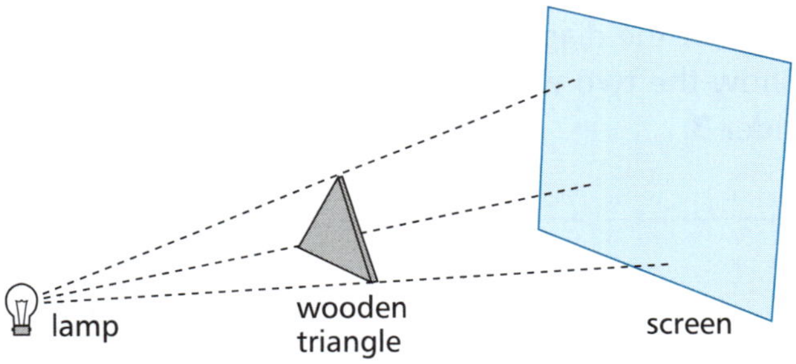

b Compare the size of the shadow with the wooden triangle.

____

____

Question 7

a Label the diagram with the following sentences.

1. Your ear collects the sound vibrations in the air.
2. Your eardrum vibrates.
3. Vibrations pass onto small bones inside the ear.
4. The vibrations in the air hit the eardrum.
5. Your inner ear sends a message to your brain via the nerves.

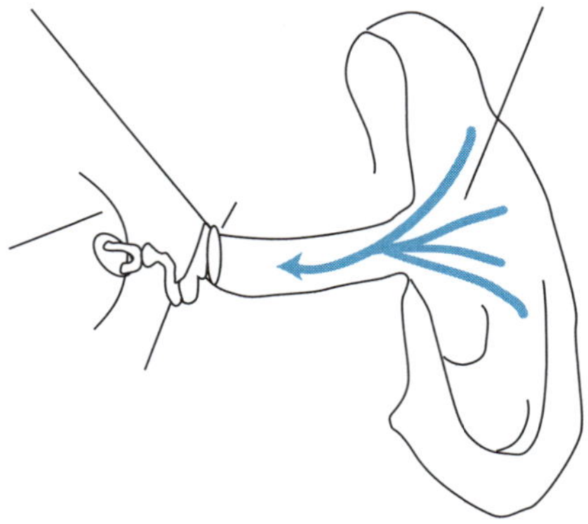

b List the sentences in the correct order.

____

Question 8

Below is a model of the water cycle.

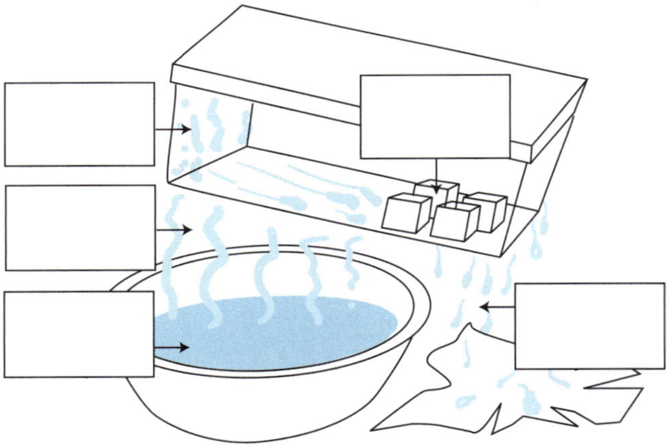

Fill in the boxes with the correct words given below.

water evaporates, ice cools the air, water condenses, drops of water fall, warm water, air cools

Question 9

Write the energy transfer sequences below, using the following words:

chemical, heat, electrical, light, kinetic

a

b

Question 10

a The two cans below were filled with hot water. Which can cools down more quickly? Which can is losing heat more quickly?

____

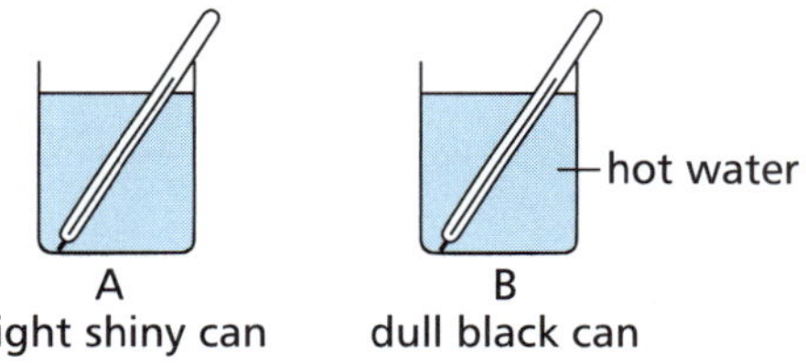

b The two cans below are filled with cold water. Which can heats up more quickly? Which can is absorbing energy more quickly?

____

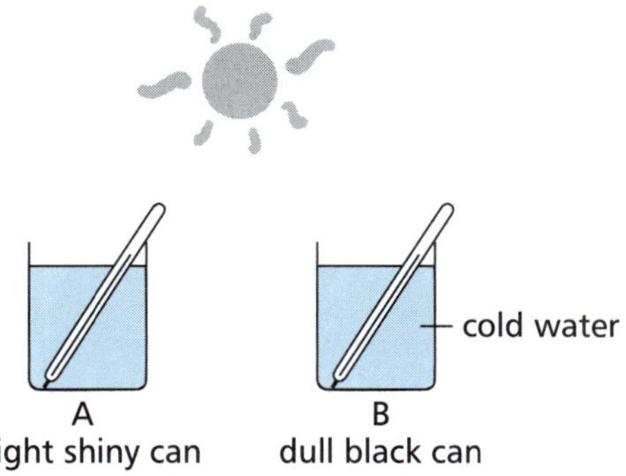

Revision questions

Question 1

Fill in the missing words.

a We measure temperature with a ______________.

b If there is a difference in temperature between the ends of a metal rod, then ______________ will flow from the ______________ end to the ______________ end.

c Convection occurs more effectively in liquids and gases than in solids because the ______________ can move more easily in them.

d ______________ occurs more effectively in metals than other solids.

e Light-coloured clothing keeps you cool on sunny days because it can ______________ radiant energy.

Question 2

A tin can is used to build the object below. By rolling the can, the elastic band winds up. Explain what happens to the can, in terms of energy transfer, when it is let go.

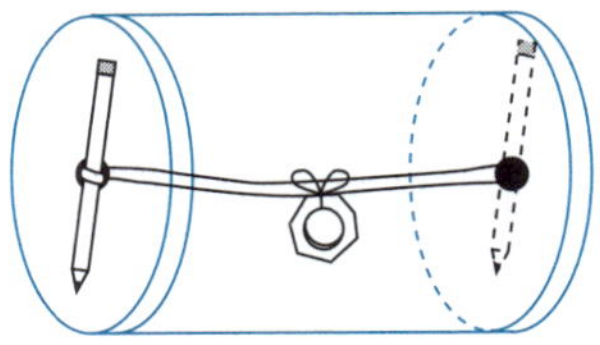

____

____

____

____

Question 3

In daylight wattle flowers look yellow.

a Why do wattle flowers look yellow?

b Why do the stems and leaves look green?

Question 4

Raindrops split white light into all the colours of the rainbow.

If we spin the coloured disc opposite, what colour will it appear to be?

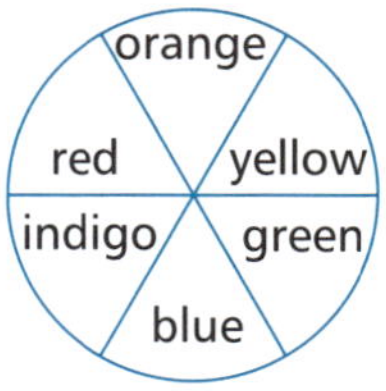

Question 5

a Why is it not as light on a cloudy day as it is on a sunny day.

b You often cannot see the sun or the sky on a cloudy day. What else can't you see on a cloudy day?

Question 6

Fill in the missing words.

You can only see when there is some ________. The things you see either give out light or ________ light into your eyes. Light travels in ________ ________, so it cannot go around ________. When light cannot pass through something it makes a ________.

Question 7

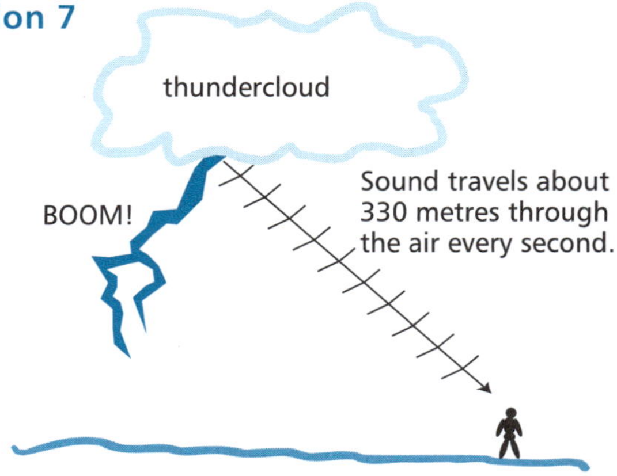

a Why do you see lightning before you hear thunder?

b You see a flash of lightning and 4 seconds later you hear thunder. How far away was the lightning?
Clue — Sound travels about 330 metres through air per minute.

Question 8

a Label the circuit below with the following words: battery, copper wires, bulb, switch.

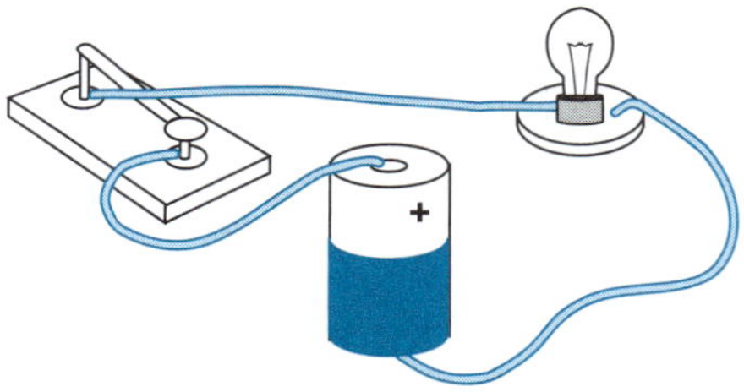

b Draw in the current flow with arrows.

c How do you know if an electric current is flowing through the circuit?

d If the copper wire was changed to string would current flow?

e What do we call a material through which electricity does not flow.

f What must you do to switch off the current?

Question 9

Look at the diagrams and write down three reasons why we need to transfer energy from fuels.

____

____

____

Question 10

a How does your body get the energy it needs?

____

b What two types of food supply most of your energy?

____

c The food we eat is stored as what kind of energy?

____

Question 11

What type of energy (potential or kinetic) do the following pictures demonstrate?

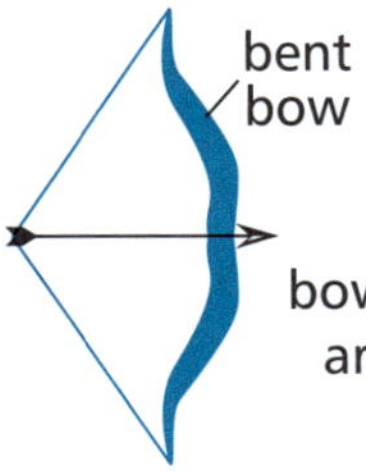

a
bow and arrow

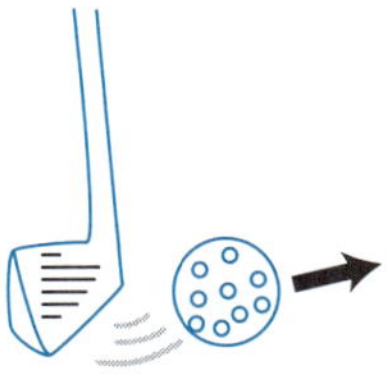

b
The ball moves.

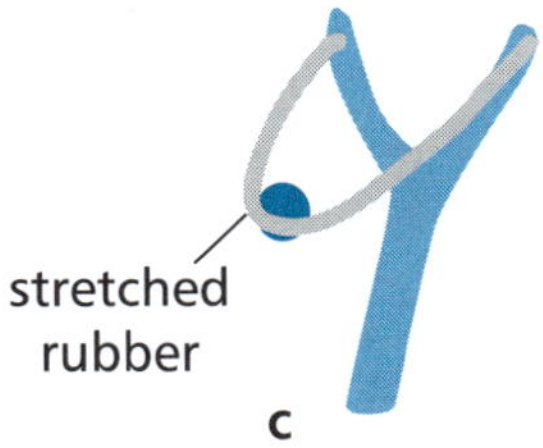

c
catapult

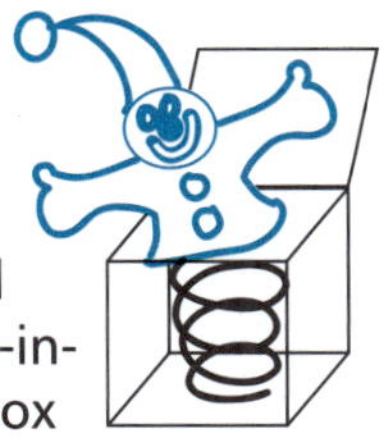

d
Jack-in-a-box

Question 12

This diagram explains the energy transfers that occur when petrol is burned in a car.

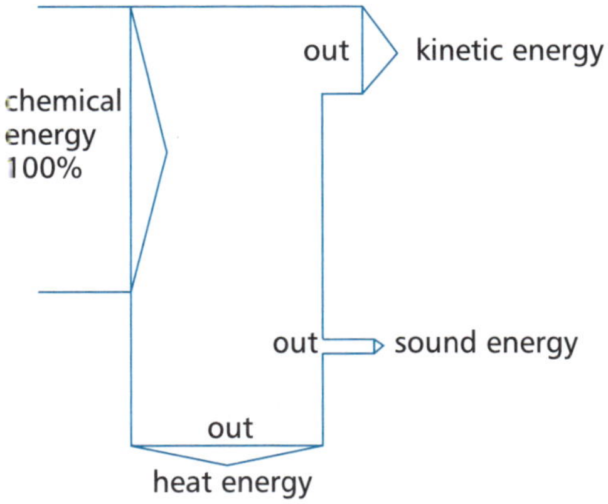

Energy In	=	Energy Out		
Energy into the engine from fuel		kinetic + energy	heat + energy	sound energy
100%		____% +	____% +	____%

Question 13

Write down the energy transfers involved in blowing up and bursting a balloon.

a Blowing the balloon up.

____

____

b Fully expanded balloon.

____

____

c Bursting balloon.

____

____

Question 14

Which form of energy is needed for the activities below?

electrical, heat, kinetic, potential

a cleaning your teeth ____________

b using a CD player ____________

c toasting a sandwich ____________

d water running out of a tap ____________

Question 15

Match the objects below to the correct energy change.

Object	Main energy change
torch	nuclear to electrical
battery	chemical to heat and light
lift going up	electrical to light
solar cell	chemical to electrical
TV	electrical to kinetic
bonfire	light to electrical
nuclear power plant	electrical to light and sound

Question 16

Students put cold water and some paper dots from the paper punch into a coffee pot.

a If the pot is heated what happens to the paper dots?

b What form of energy transfer is this movement called?

Question 17

What energy changes are happening in the cartoon opposite? Write them in the boxes.

Question 18

Write down the names of:

a Three musical instruments that use vibrating strings to make sounds.

b Three things you hit to make sounds.

c Three instruments that use vibrating air to make sounds.

Question 19

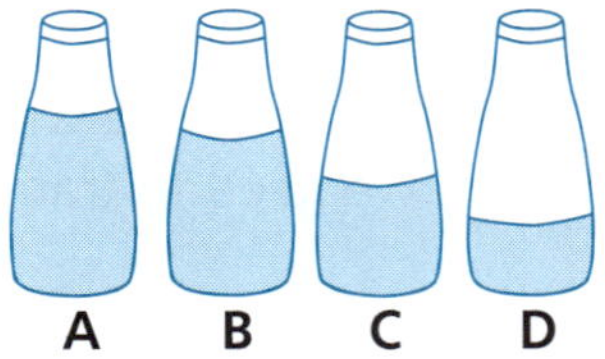

If these bottles are tapped with a wooden spoon —

a Which milk bottle will make the highest sound?

b Why do they make different sounds?

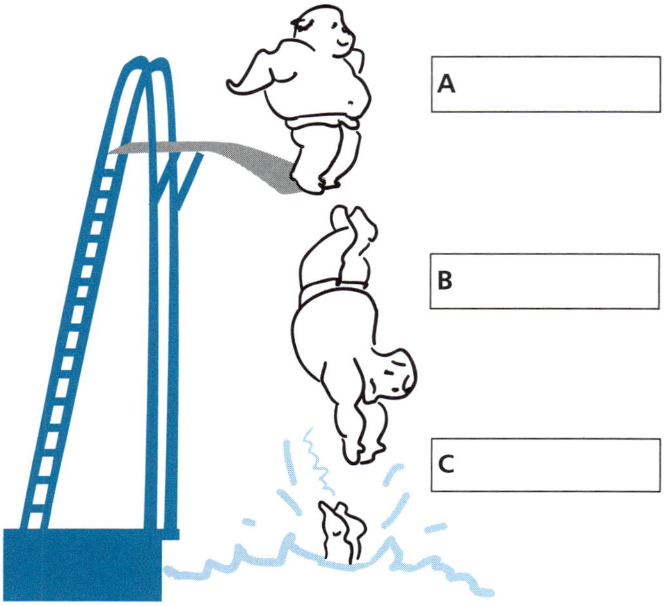

Question 20

Using the equipment listed below design a solar hot water system.

Equipment:

black plastic hose
plastic bottle
insulating material
cardboard box
water
thermometer

CHAPTER 8

ENERGY (SECTION 2)

Did the big dinosaurs roar or squeak?

Light

We can see things because they give out their own light or because they reflect light that comes from something else.

Daytime — During the day the sun's light lets you see shapes and colours. Sunlight bounces off things (is reflected) and your eyes see the reflected light.

Nightime — At night there is not enough light to see things clearly. People have always looked for ways to light up the dark — firelight, candlelight, oil lamps, gaslight, electric lights.

Bending light rays

Light travels in straight lines but you can make it change course. If light travels through different substances it will bend at the boundary between the substances. This bending is called **refraction**.

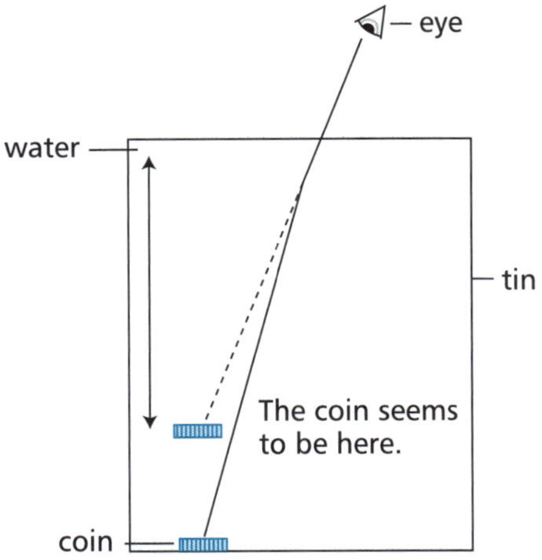

When light crosses the boundary between water and air it is refracted. It is refracted away from the 'normal' — the line at right angles to the water surface.

Water into Air

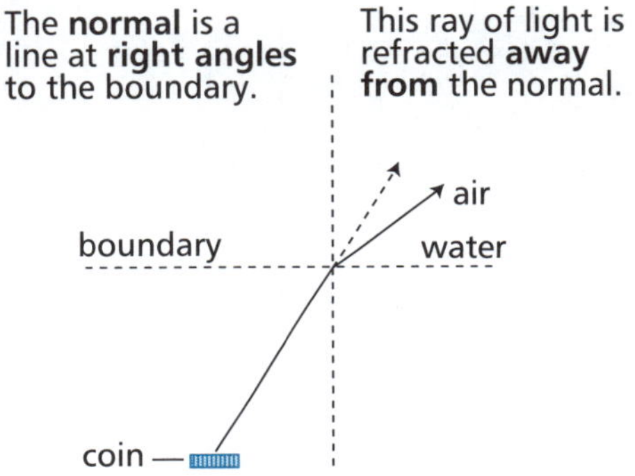

Air into Water

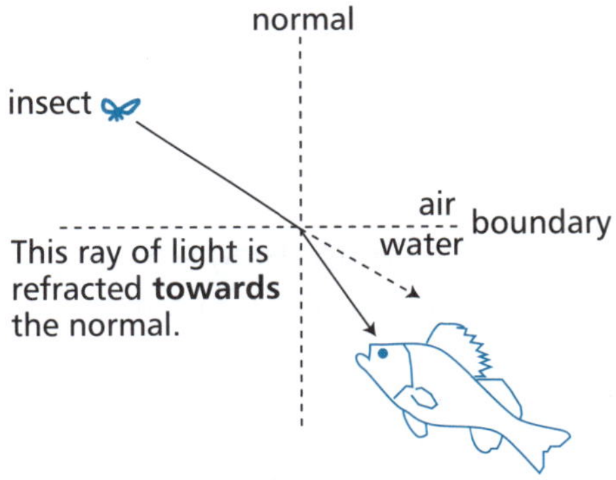

When a ray of light passes from water to air it is refracted and makes things look different.

Water looks shallower than it really is. It seems to be three-quarters as deep as it really is. Anything under water also looks smaller.

A glass prism will bend light and in so doing breaks it up into the rainbow colours. Light at the red end is not **refracted** as much as the light at the violet end of the spectrum.

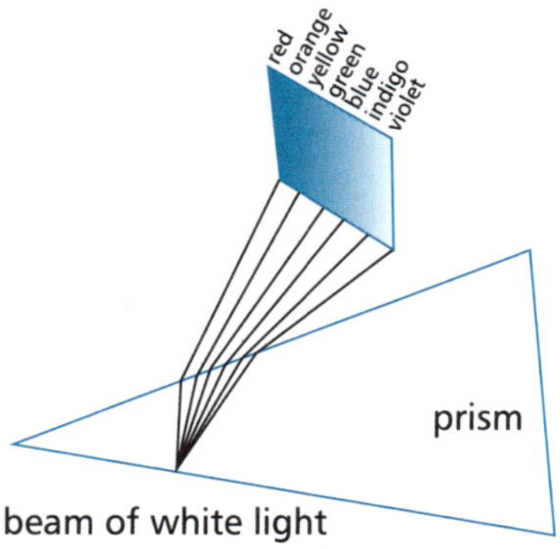

Splitting white light using a prism

The colours can be mixed together again by using another prism.

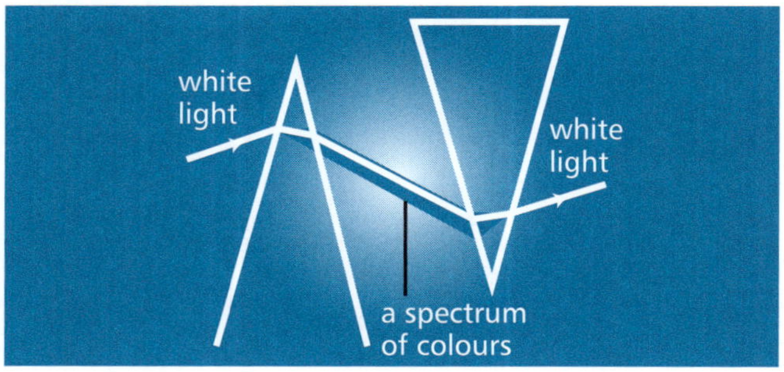

Prisms don't always split white light. They can be used to send light in a different direction.

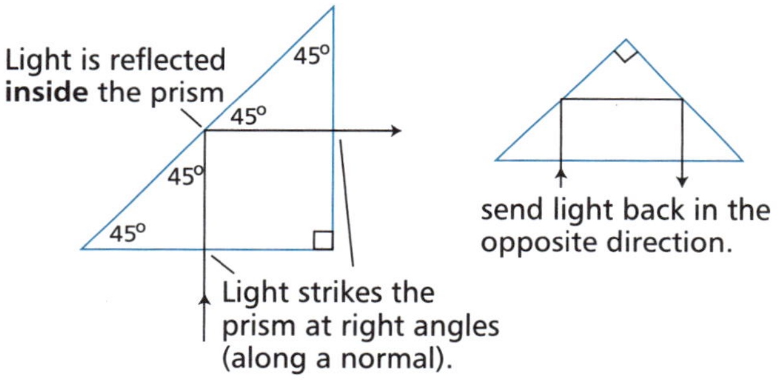

This system is similar to using mirrors and is used in some periscopes and in binoculars.

Comparing mirrors and prisms

If you put two mirrors together you can also change the direction of light.

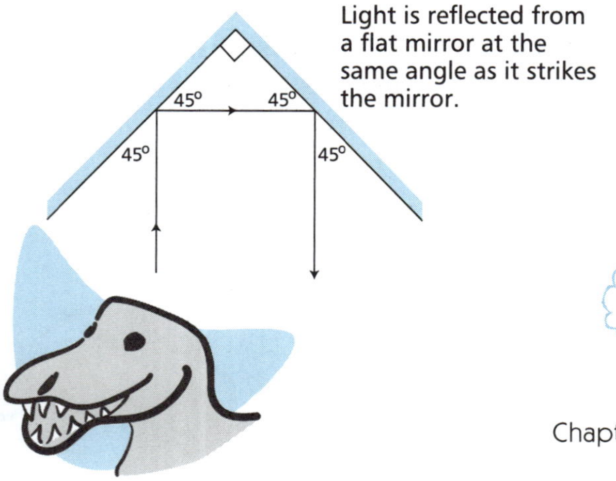

Curved mirrors

Have you ever seen your image in a spoon?

A **concave** mirror is a converging mirror. And a **convex** mirror is a diverging mirror.

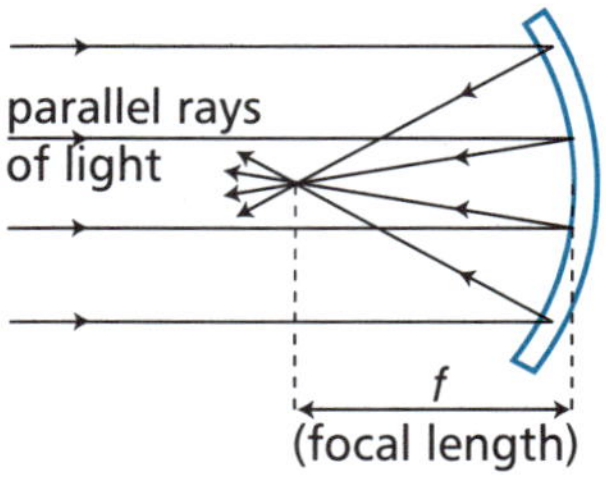

Concave mirror

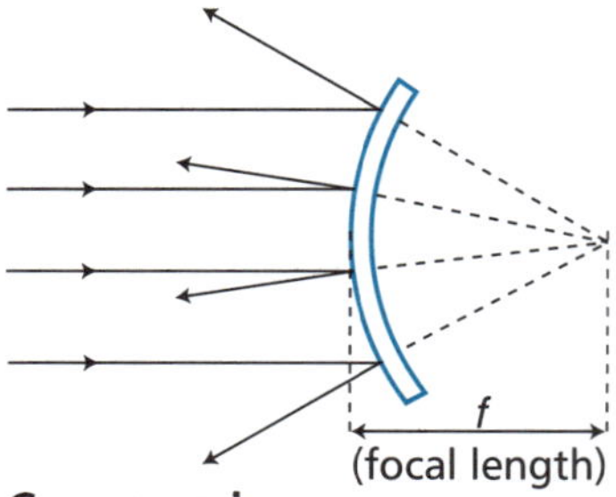

Convex mirror

What about lenses?

Curved lenses can cause light rays to spread out (**diverge**) or come together (**converge**).

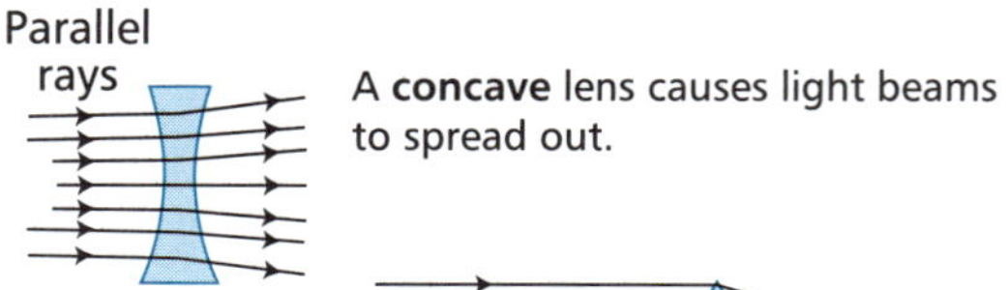

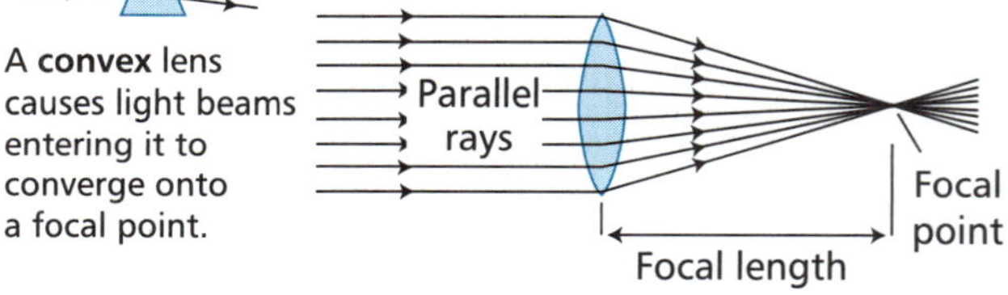

The camera and the human eye work much the same way. They both focus light through a curved lens. The diagrams below give a basic comparison of the eye and the camera.

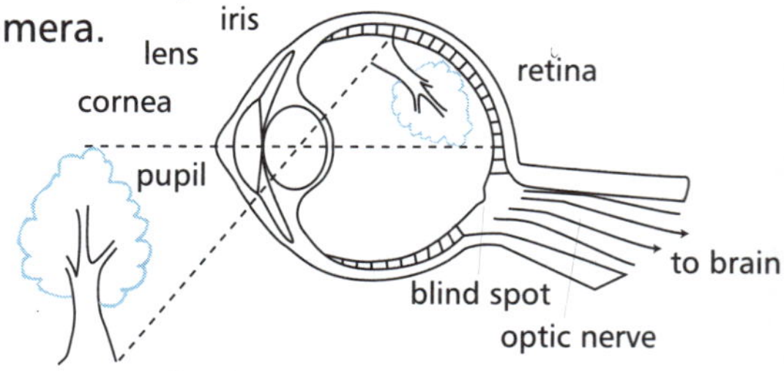

The human eye

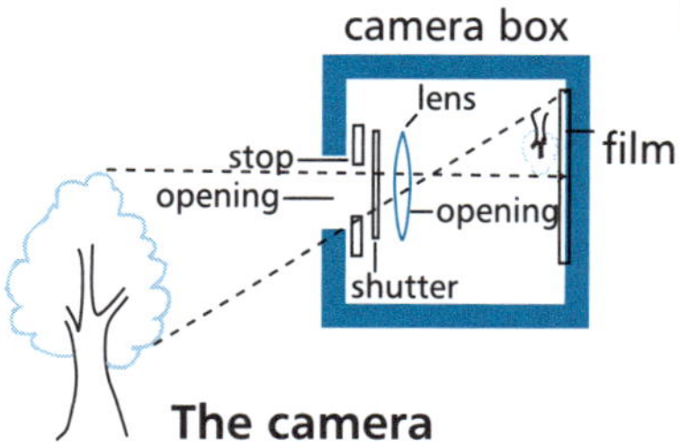

The camera

The shutter and the iris open and close to control the amount of light. The image from the lens forms on the retina in the eye or the film in the camera. In the human eye the brain turns the image up the right way.

Sound

- All sounds are made by vibrations.
- Vibrations can travel through solids, liquids and gases but they cannot travel through empty space.
- Sound travels slower than light.
- Quiet sounds have small vibrations and loud sounds have large vibrations.
- We say loud sounds have a big amplitude.
- Loudness (noise level) is measured in decibels (dB) using a sound meter.
- The number of vibrations each second is called the **frequency**. Vibrations that have a high frequency (many complete to-and-fro movements per second) produce sounds which have a high pitch.
- Frequency is measured in hertz (Hz).

  A guitar string that vibrates 196 times in one second has a frequency of 196 Hz.
- To produce vibrations with a higher frequency you need to:
  - tighten the string
  - shorten the string
  - use a thinner string.
- Instruments that measure vibrations are called oscilloscopes.

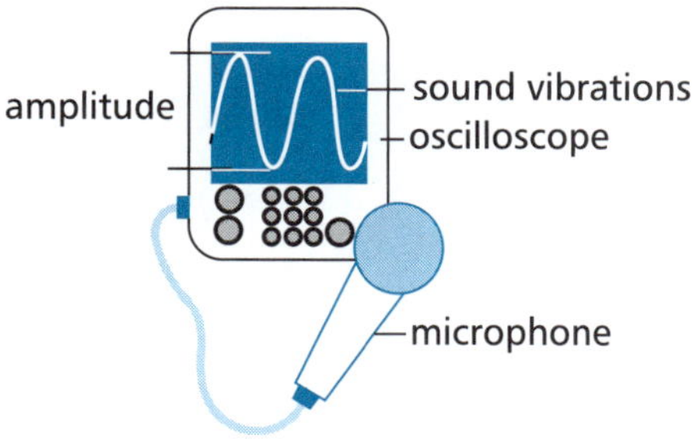

Oscilloscope with sound vibrations

Electricity

Where does electricity come from?

Basically all matter is electrical. The atoms of matter are made up of small particles called **electrons** (carrying a negative charge) and **protons** (carrying a positive charge).

We don't make electricity, we just have to make the electrons flow through a substance to make an electric current.

Two kinds of electricity

1 Static electricity

- This is electricity that does not flow through wires. Lightning is the transfer of electrons from clouds to other clouds or to the ground. When there is enough charge present electrons will jump across a gap —we see this as a spark.
- You can give an object an electric charge by rubbing it with a different material. When you use a plastic comb, it rubs against your hair. The comb becomes charged with electricity (either a positive or a negative charge) and will pick up small bits of paper or even your hair.
- Two objects that have electric charges either attract or repel each other. Two positive or two negative charges repel each other but a positive and a negative charge attract each other.

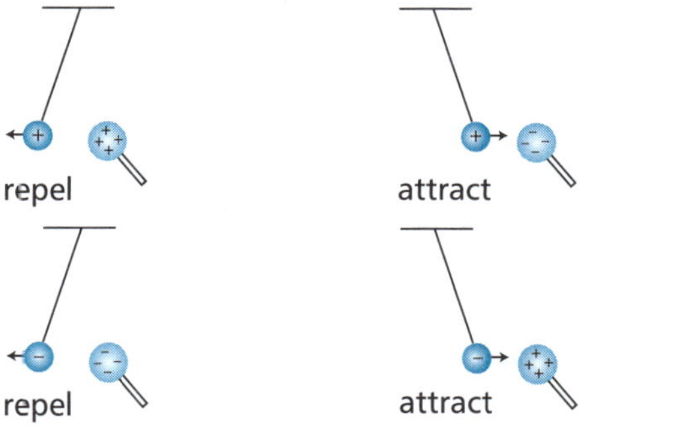

Like charges repel **Unlike charges attract**

- We cannot use static electricity to power electric appliances because we cannot control the flow of electrons. Current electricity does let us control the flow of electrons.

2 Current electricity

- When charges move they make an electric current. Current electricity will flow through wires and do lots of useful jobs.
- Electric current comes from the mains (you can plug it in and turn it on) or from batteries (safest).
- To get an electric current from a battery you need to build a circuit.

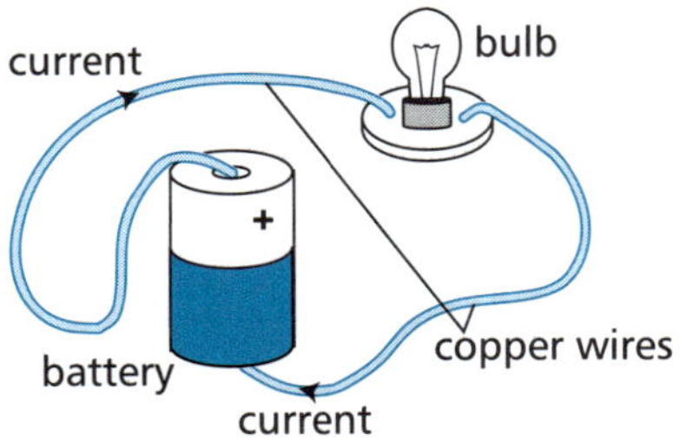

Circuit

- Electrons flow through some substances easily (**conductors**) and not so easily through others (**insulators**).
- An electric circuit can flow only if the circuit is made of conductors.

Building circuits

The simplest circuit is a bulb and a battery.

If you add a switch it is more useful.

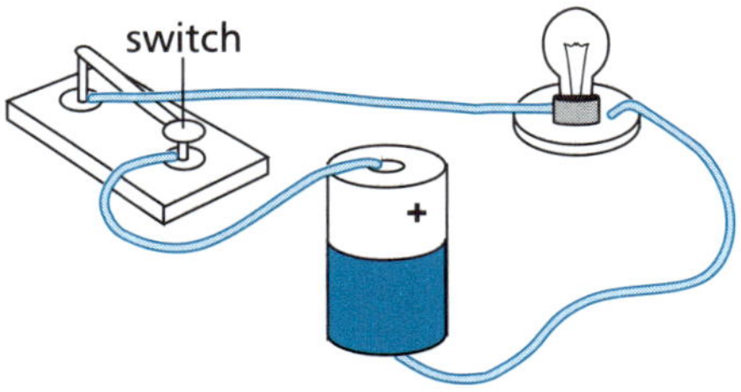

Circuit with a switch

We could add another bulb to the circuit so that current flows through one bulb and then through the other. Bulbs connected in **series** share the current flowing in the wire so they are both dimmer.

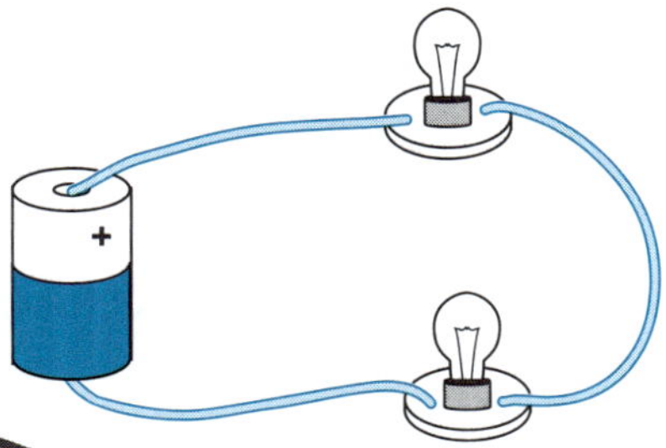

Bulbs connected in series

If we connect up two bulbs from the same battery but on separate wires, the bulbs are connected in parallel and they are both very bright because they are not sharing the electrons flowing through the wire. They each have the same supply.

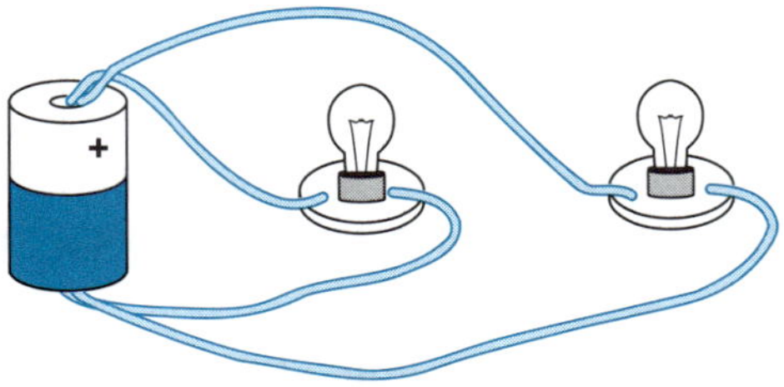

Bulbs connected in parallel

Circuits can be drawn using symbols.

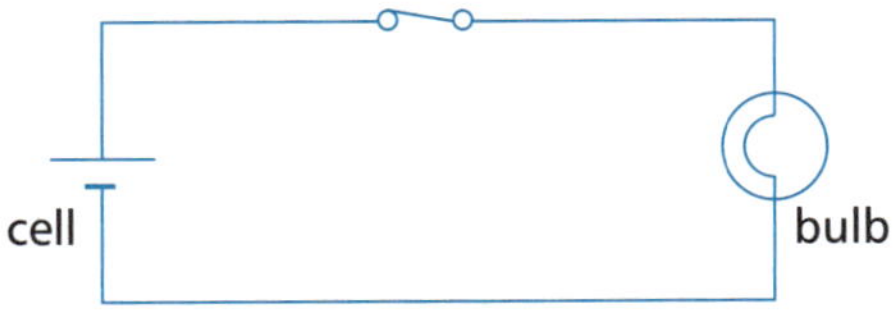

If the switch is off, you should draw it like this

In a parallel circuit the switch could switch off both bulbs or just one bulb depending on where the switch is located.

The filament of a bulb gives out light when the electrons pass through it. If the filament breaks this stops the flow of electrons.

Structure of a bulb

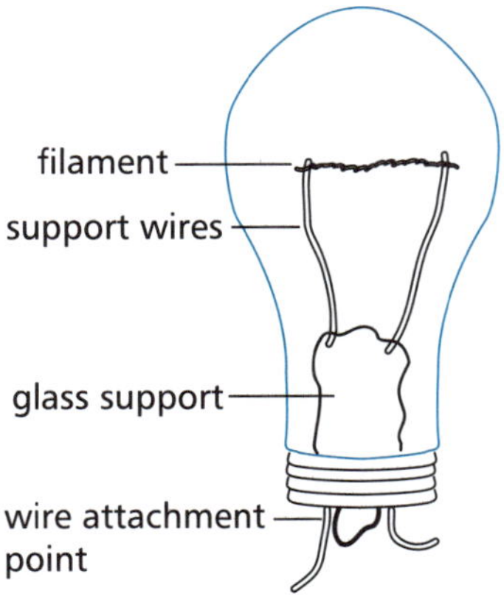

A fuse or fuse wire is designed to break and stop the electron (current) flow if too much current is flowing. Too much current can cause house fires.

Glossary

Amplitude
The size of the vibrations that produce sounds. A small amplitude produces a small sound.

Circuit
A pathway through which electrons (current) flow.

Concave
A mirror that curves in like a cave.

Conductor
A material through which electrical charges flow easily.

Converge
Light rays come together.

Convex
A mirror that curves outwards.

Current
Electric charges flowing through a wire in a complete circuit.

Diverge
Light rays are spread out.

Electrons
The negatively charged particles in an atom.

Frequency
The number of vibrations per second. This gives a sound a particular pitch.

Hertz (Hz)
Unit for frequency — number of vibrations per second.

Insulator
A substance that will not let an electric current pass through it.

Iris
The coloured part of the eye that opens and closes to control the amount of light reaching the retina.

Lens
Collects light from an object and focuses it to form a sharp image.

Pitch
Pitch can be high or low depending on the frequency of the vibrations. A roar has a low pitch and a squeak has a high pitch.

Protons
The positively charged particles in an atom.

Refraction
The change of direction of a ray of light in passing from one substance to another.

Retina
The image from the lens forms on this part of the eye.

Question 1

Draw the path of light through the following diagram of a periscope.

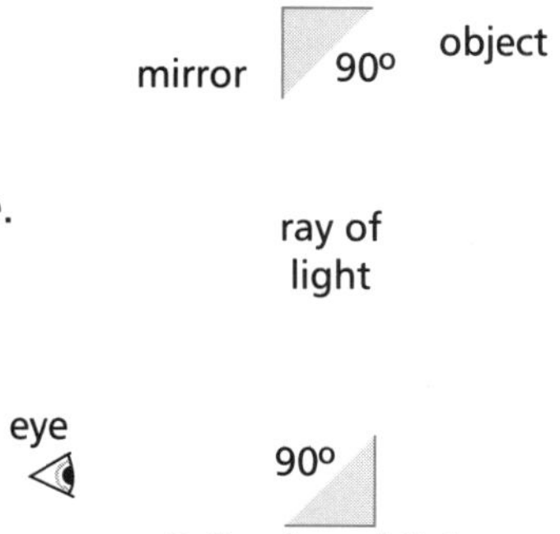

Reflection of light in the periscope

Question 2

In the diagrams below draw the path of light rays as they pass from the left to the right through the lenses.

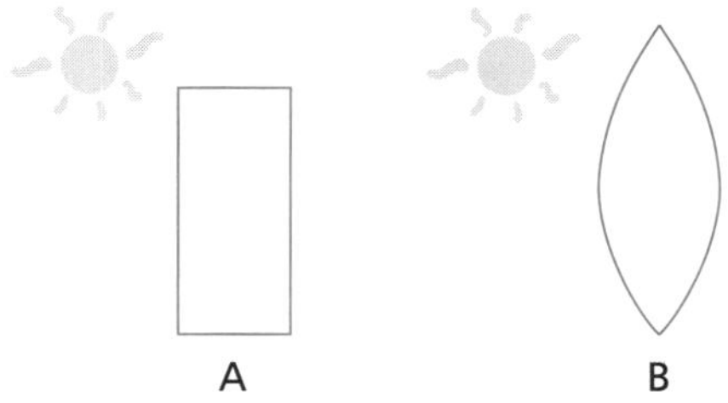

Question 3

Which of the following drawings describes how you see a tree?

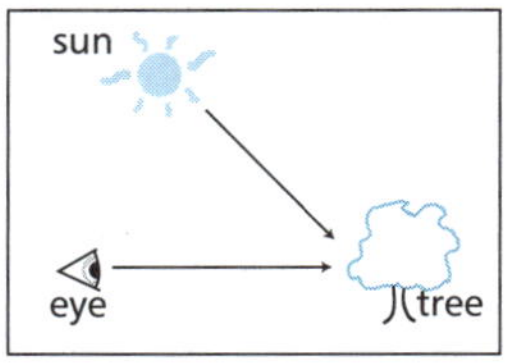

a Light goes to tree and we look at tree.

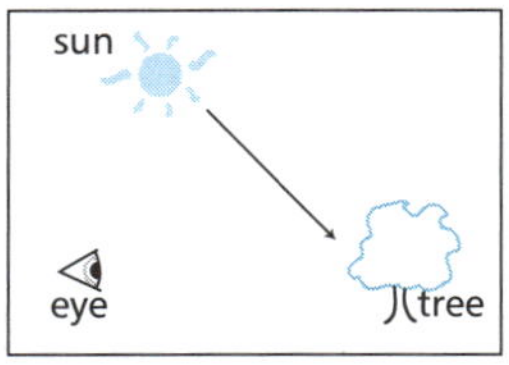

b Light shines on tree and we see it.

c Light is everywhere and we can see it.

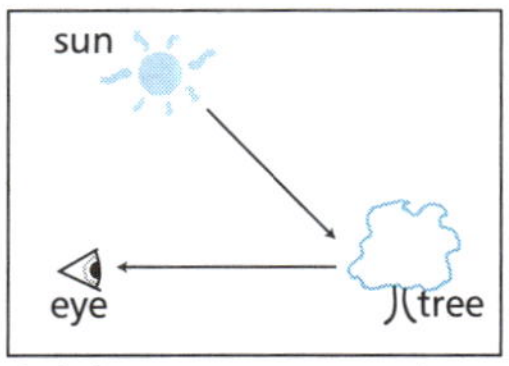

d Light goes to tree and bounces to our eyes.

Question 4

Explain what is happening in the following photo.

Question 5

The diagram shows a dish-aerial which is used to receive television signals from a satellite.

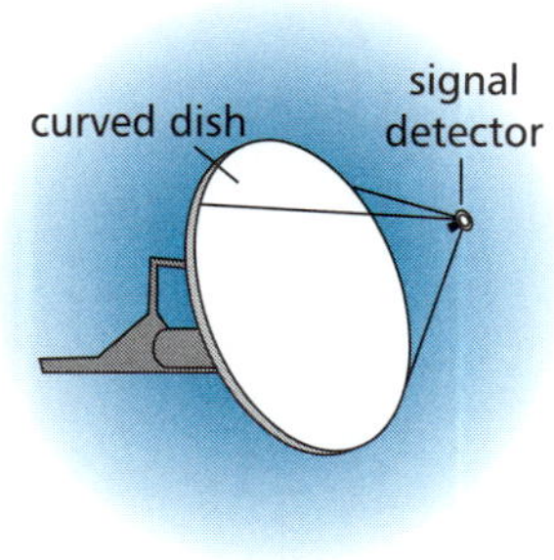

a What is the dish used for?

b Should it be concave or convex?

c Where should the aerial be positioned to receive the strongest possible signal?

d What change would you predict if a larger dish was used?

Question 6

Students set this experiment up in a dark room. The tin has a hole in the bottom and is filled with water. They shined the torch into the tin and observed the light travelling along the spout of water.

Draw the path of light in this diagram and explain this in terms of reflection of light.

Question 7

Light rays bend when they pass from one substance to another.

a What scientific term is used to describe this?

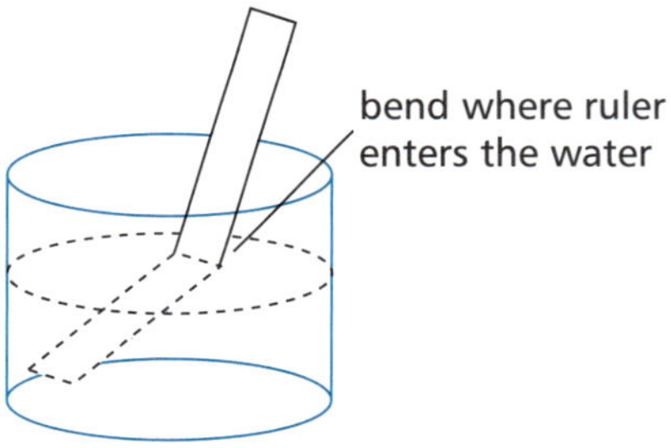

b Why does the ruler appear to bend?

c Can you explain in terms of refraction what would happen if a car travelling very fast on a straight bitumen road unexpectedly hit a muddy patch. (Mud will make the car slow down and swerve.)

Question 8

a Study the following table and tick the boxes where the light ray is reflected off a smooth mirror.

Angle that light strikes the substance	Angle that light is reflected
20	23
45	45
23	23
67	59
87	87

b Fill in the missing words.

A smooth mirror ____________ a ray of light at the ____________ angle as it strikes the mirror.

Question 9

Some sounds are louder than others.

Compare the following diagrams of a guitar string.

A String plucked gently:

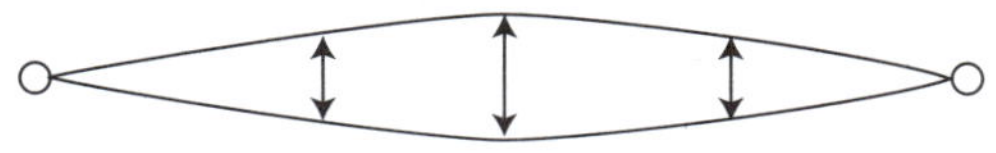

B String plucked harder:

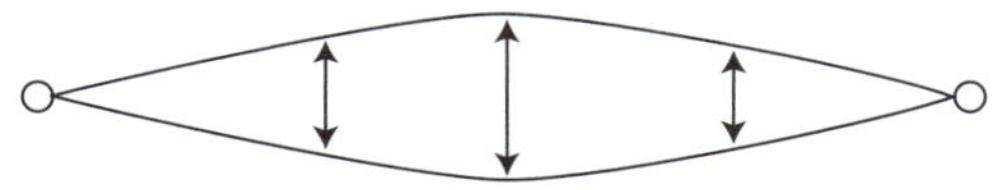

a Which string will produce small vibrations and a quiet sound?

____

b What is a way of saying that the vibrations are large.

____

Question 10

The pitch of a sound is how high or low the sound is. Frequency is how many vibrations there are each second, measured in **hertz**.

Study the diagram and answer the questions.

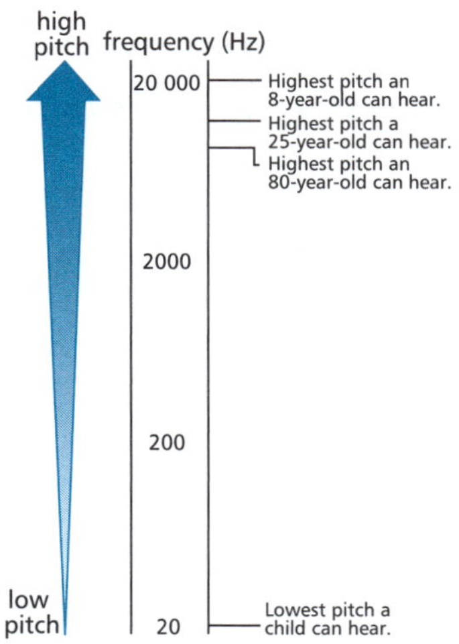

a What is the lowest frequency that you can hear?

____

b What happens to the highest frequency people can hear as they get older?

____

c The following graph represents the hearing ability of a child. At which frequencies does he have the most serious hearing loss?

____

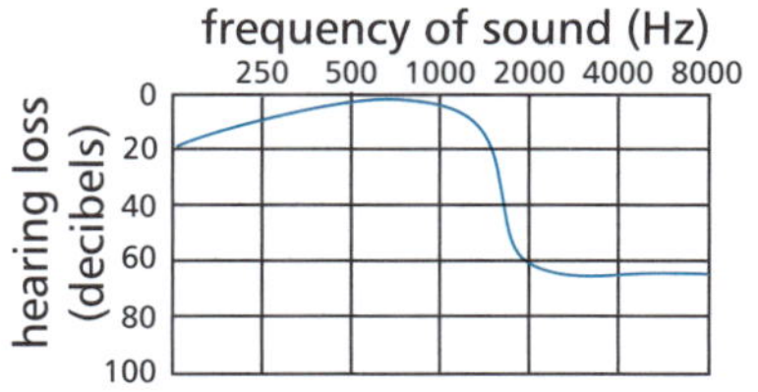

Question 11

An oscilloscope measures vibrations.

Look at the waves on the screens caused by a high pitched tuning fork and a low pitched tuning fork. Which is the high pitched vibration?

____

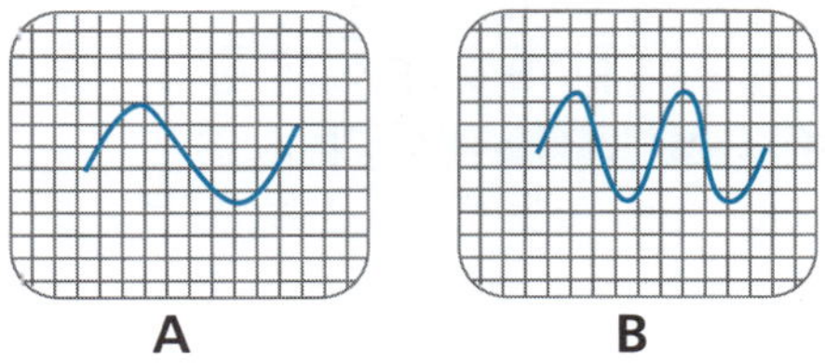

Question 12

These instruments make a sound because the air inside vibrates.

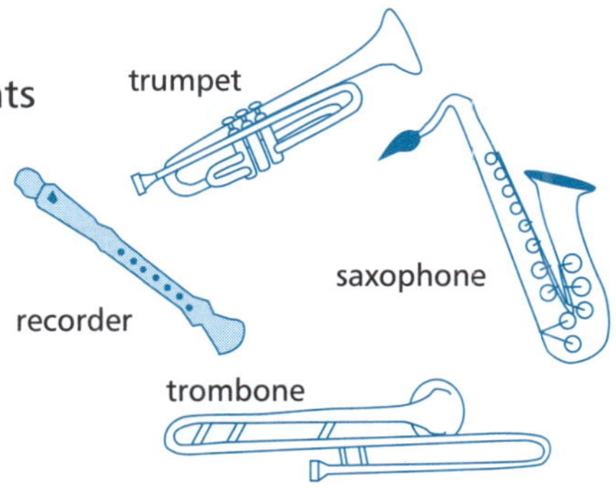

Explain the relationship between the pitch (frequency) and the length of the pipe?

____

____

____

Question 13

A sound meter is used to measure the noise level in decibels (dB) at different distances from the loudspeaker at music festival.

Distance (m)	1	2	3	4	5
Noise level (dB)	150	140	130	120	110

a Plot a line graph and extend it to find the noise level at 6 m.

b Look at the scale below. Is the level shown above, safe for our ears?

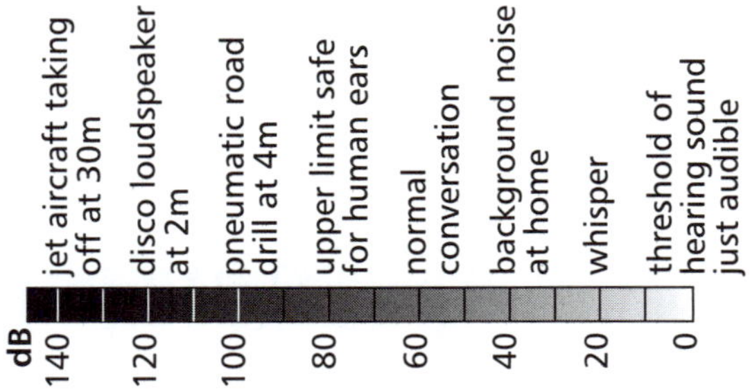

Question 14

Geologists use echo-sounding to search for minerals. Sound travels through rock at 4000 m/s.

a Which microphone will receive the sound first?

b If it takes 0.5 seconds for sound to arrive at A, how deep is the mineral deposit?

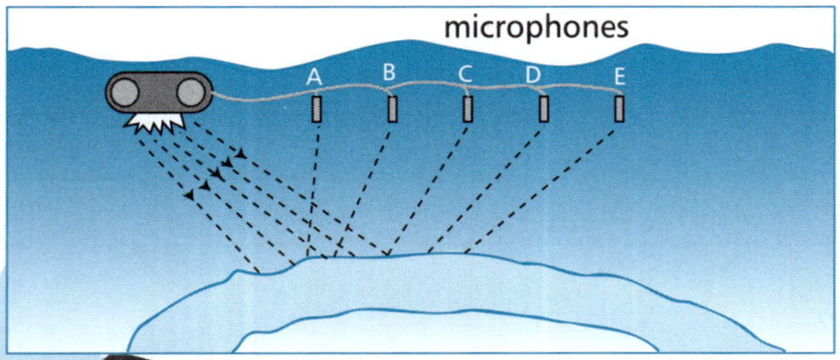

Question 15

Predict which bulbs will light up.

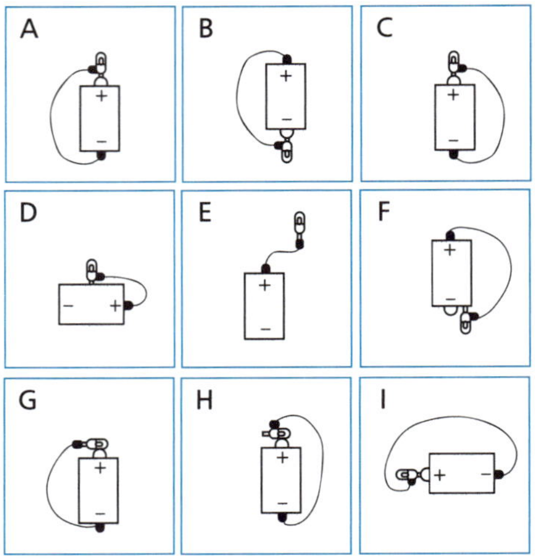

Question 16

A battery is connected up to a bulb and the bulb is glowing.

Which of the following diagrams best describes the electric current in the wires?

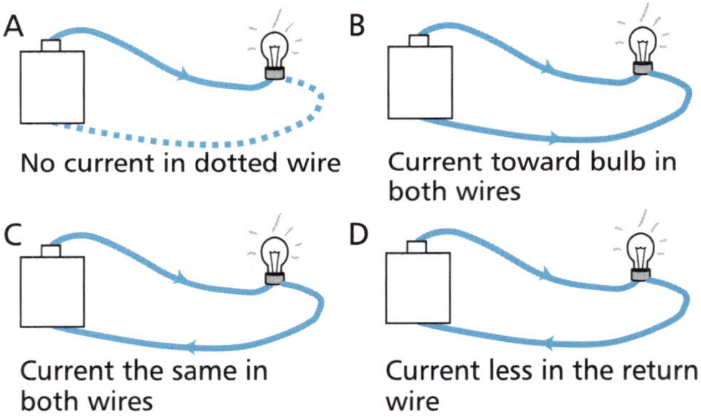

Question 17

The bulbs in the following three circuits are connected differently.

Match each circuit with the correct description:

Connected in series or **connected in parallel**

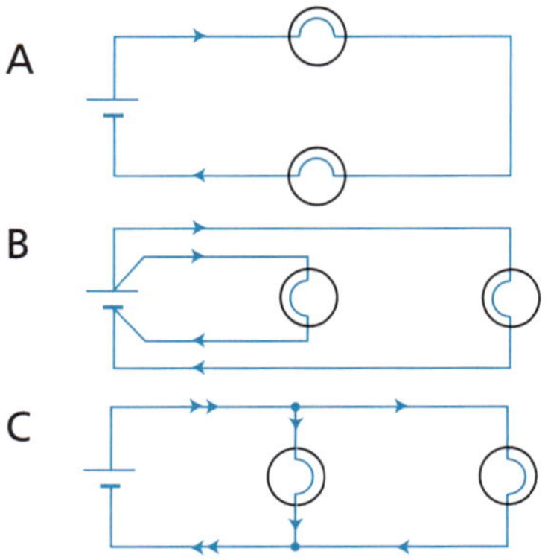

Question 18

Look at the two circuit diagrams below.

Which bulb does the switch turn on in each case?

____

____

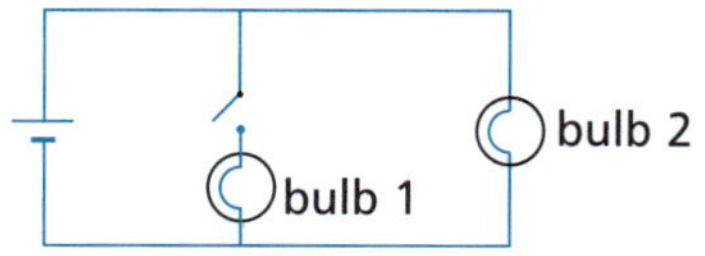

Circuit A

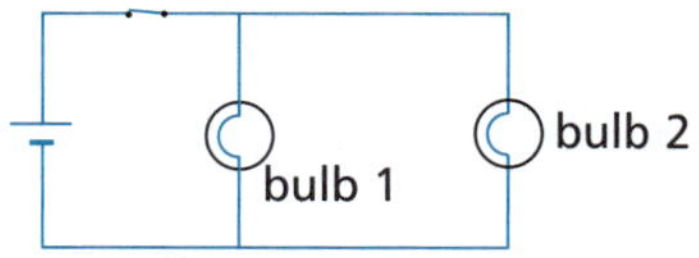

Circuit B

Question 19

Neither of the bulbs in this circuit light up.

a Why doesn't bulb A light up?

____

b Why doesn't bulb B light up?

____

____

____

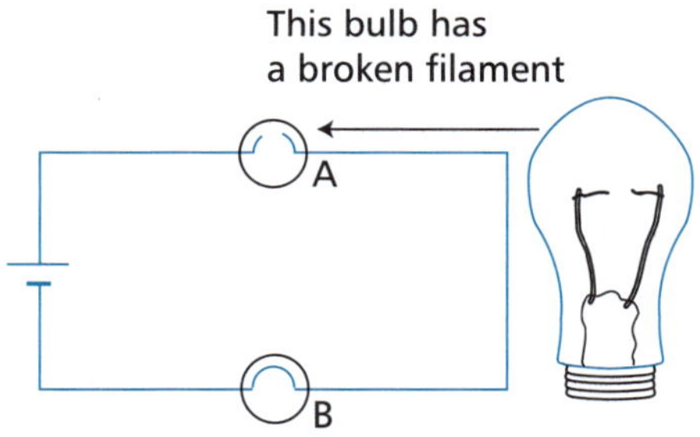

Question 20

Complete the following table.

Charge on object 1	Charge on object 2	Do they attract or repel?
−	−	
+	−	
+	+	

Question 1

a Insert the correct words.

Water always looks ____________ than it really is. A ruler standing in a container of water looks ____________. These things happen because light is ____________ away from the ____________ when it passes from water into ____________.

b Answer the following questions:

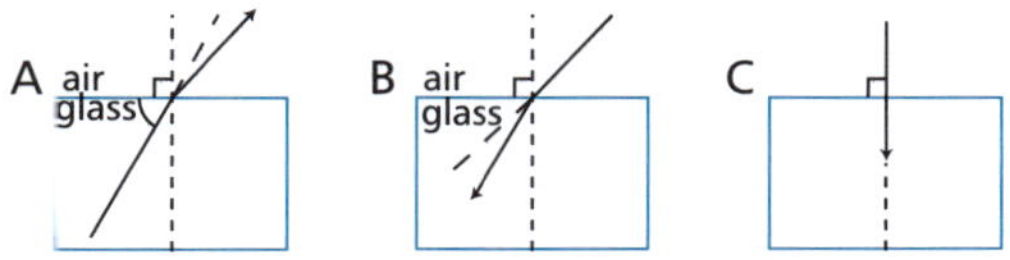

A When light passes from glass into air at 50° to the boundary it is refracted ____________ from the normal.

B When light passes from air into glass or water it is refracted ____________ the normal.

C When light crosses a boundary at ____________ (90°) it is not refracted.

Question 2

a The lanes on a road are marked with cat's eyes. Explain how these work.

____

____

____

b Why does a bicycle reflector appear to be red?

____

____

____

Question 3

Drivers use mirrors to see what is behind them.

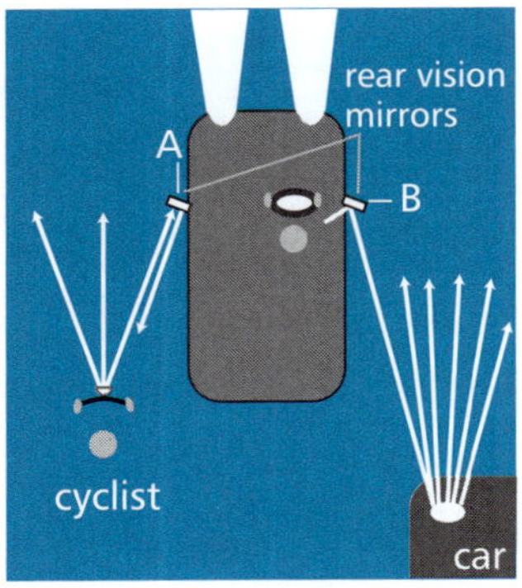

a Which mirror does the driver use to see the car that is overtaking her?

____

b Why can't the driver see the cyclist on her left?

____

____

c What must the driver do so that she can see the cyclist?

____

Question 4

a Label the diagram below with the following words: iris, pupil, retina, optic nerve, cornea, lens, object

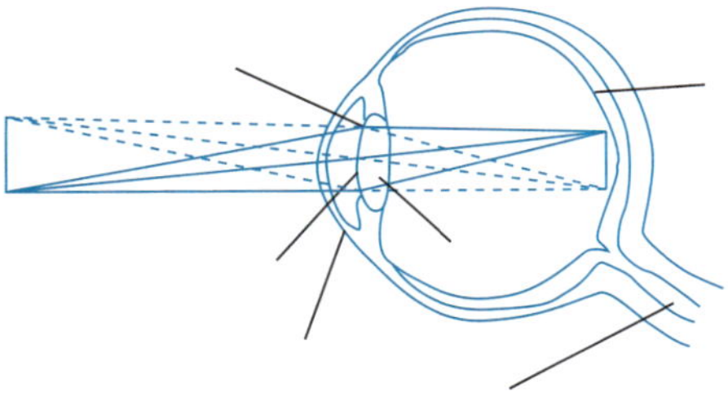

b Match the words with the statements on the right.

retina	a controls the pupil diameter
pupil	b focuses objects on the retina
optic nerve	c diameter decreases in bright light
lens	d carries nerve impulses from the retina to the brain
iris	e location of the visual receptors

Question 5

Could the candle flame be seen if the position of one of the cards is moved? Why?

____

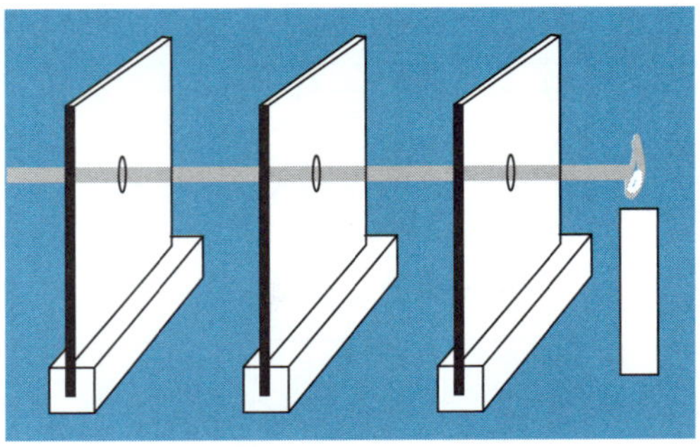

Question 6

Which of the surfaces below is the best reflector?

____

____

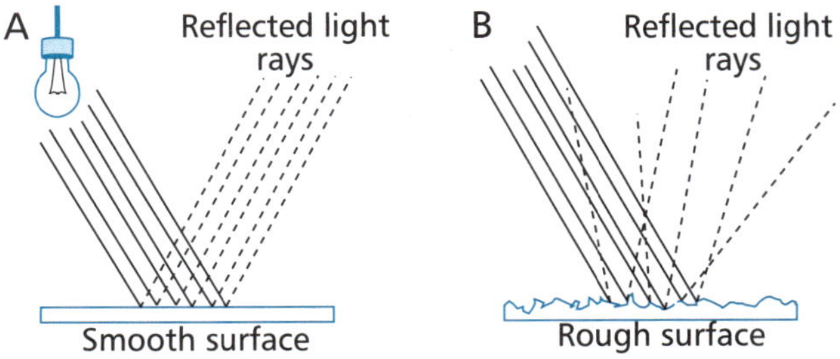

Question 7

Correctly label the colours of the spectrum in the diagram below.

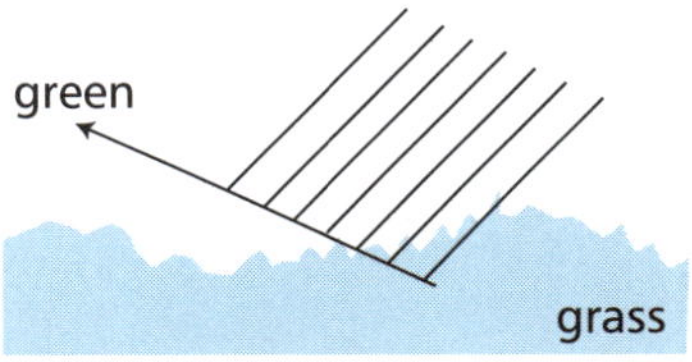

Question 8

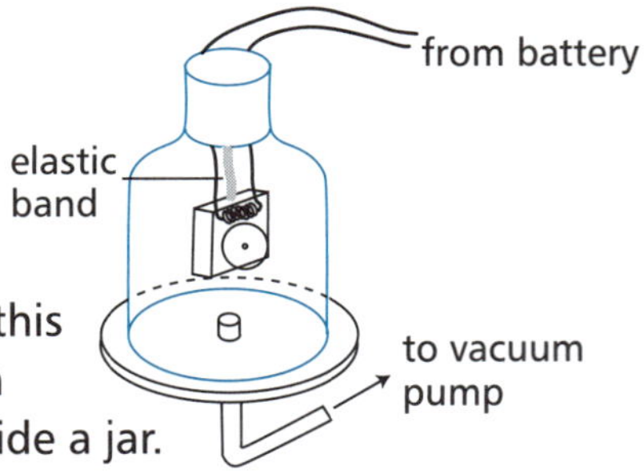

A student set up this experiment which involves a bell inside a jar.

If he switches on the bell does it ring? __________

Can he hear it? ______________________________

Explain why. ____________________

____

____

Question 9

Describe how the sound from a tuning fork reaches the ear of the listener?

____

____

____

Question 10

a Which plays notes of a higher pitch, a violin or a cello?

____

b Where do you find the lowest notes on a keyboard?

____

c Is there a difference between the strings at the left-hand end of a piano compared to the right-hand end?

____

d Are the strings of a piano scraped, tapped or plucked to make them vibrate?

____

Question 11

Loud sounds can be very dangerous.

a Why is this so?

____

____

b How can people prevent damage to their ears by loud sounds?

____

____

Question 12

An aeroplane flies overhead. You hear the aeroplane and look up but it is not where you are expecting it to be. Explain why.

____

____

Question 13

Choose from the following phrases to complete the table.

1 sound
2 light
3 both light and sound
4 neither light nor sound

a Can travel through a vacuum (space)	
b Can travel through air	
c Used by whales for navigation	
d Can be bent	
e Can bounce off surfaces	

Question 14

Fill in the spaces with the word **series** or **parallel**.

If a current flows through one bulb and then another, the bulbs are connected in ____________. If two bulbs are connected separately to battery, they are connected in ____________. A switch must be in ____________ with the bulb. In a ____________ circuit if one bulb breaks, the other bulb does not work. In a ____________ circuit, if one bulb breaks the other bulb stays on.

Question 15

Why do all the Christmas tree lights go out if one bulb only is broken?

____

____

____

Question 16

a What is the difference between circuits A and B? Will current flow through the circuit in A? Give a reason why.

____

____

____

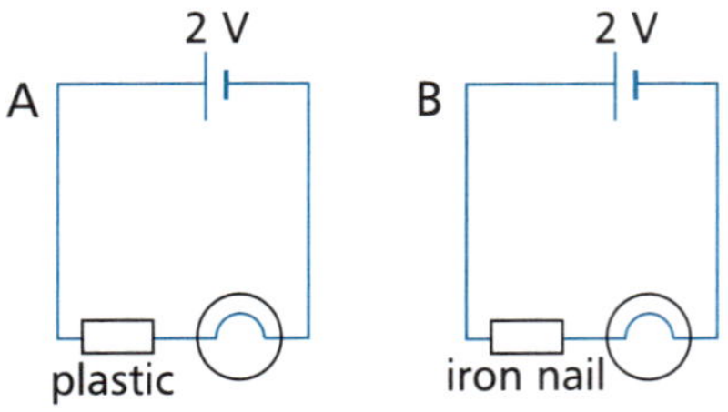

b Look at circuit C and label the statements as TRUE or FALSE.

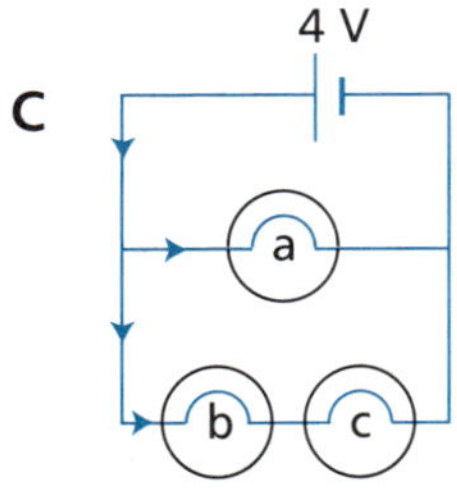

i a and b will be the same brightness

____

ii a will be brighter than b and c

____

iii b will be brighter than c

____

Question 17

a What is static electricity?

____

____

b Is static electricity useful?

____

____

c What is current electricity?

____

____

d What sorts of things do we use current electricity for?

____

____

e What is an electric circuit?

____

____

f Label the four parts of this diagram.

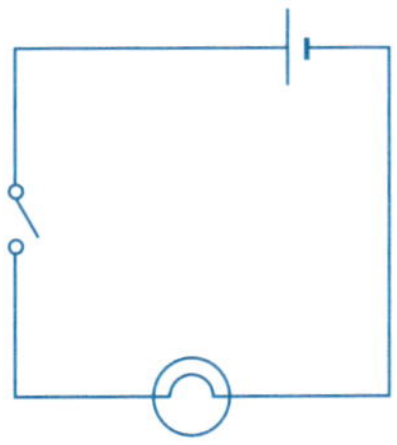

Question 18

You would like to burglar proof your room.

Design a circuit, using the following equipment, that will warn you if anyone goes into your room.

Equipment: copper wire, buzzer, battery, two drawing pins, two wooden, spring-connected, clothes pegs

Question 19

Fill in the missing words.

A ____________ is a short piece of thin wire. If too much ____________ flows through it, it overheats and melts. Fuses are put into each ____________ in the home to prevent fires.

Question 20

If we connect batteries in series, as in this diagram, how will the bulb glow compared to a circuit with only one battery?

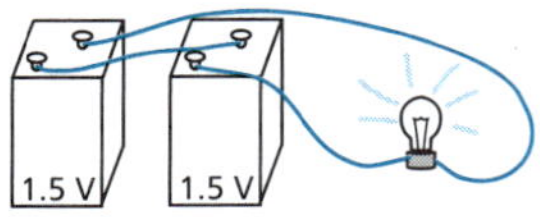

____

____

Test B

Question 1

Fill in the missing words:

Each element consists of tiny particles called ___________. An atom consists of even simpler particles called ___________ (positive charge), (uncharged) and (negative charge). The centre of the atom is called the ___________ and is composed of and ___________.

Question 2

a What is the name for a substance which cannot be split into simpler substance?

b About how many elements have been discovered on Earth?

c What two elements make up water?

d Which four elements are the most common in biological materials?

Question 3

The following information is about the structure of the Earth.

Layers of the Earth	Thickness (km)	Temp. (°C)	Features
crust	10–70	20–500	solid rock
mantle	3000	500–2000	thick liquid
outer core	2300	3000	liquid iron & nickel
inner core	1300	4000	solid

a Label the diagram below with the layers of the Earth.

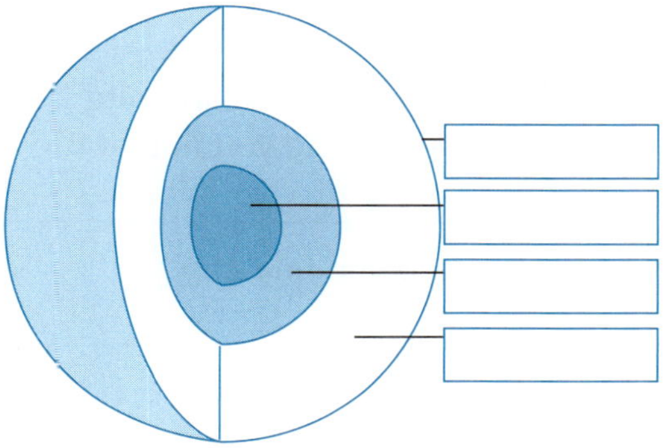

b What is the hottest part of the Earth?

c What is the thinnest part of the Earth?

d What is the relationship between the distance from the core and the temperature of the layer?

Question 4

If you wanted to find out if an igneous rock had been formed by rapid or slow cooling you would observe the:

a hardness of the rock

b shape of the rock

c colour of the rock

d size of the crystals

Question 5

Study the reaction of a volcano below and answer the questions.

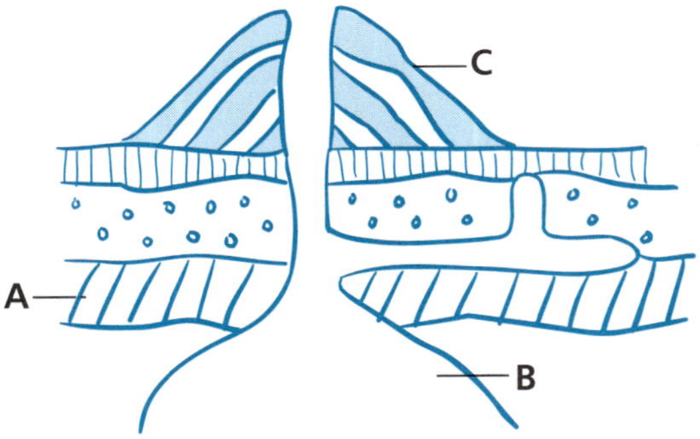

a Where would you expect to find igneous rocks with the smallest crystals?

____

b Which of the following rocks (granite sandstone, basalt) would you find at B?

____

c Which rock would you find at A?

____

d Which rock would you find at C?

____

Question 6

Three geologists drilled holes to measure the temperature of the Earth. Only one recorded the data correctly. Which geologist was correct?

____

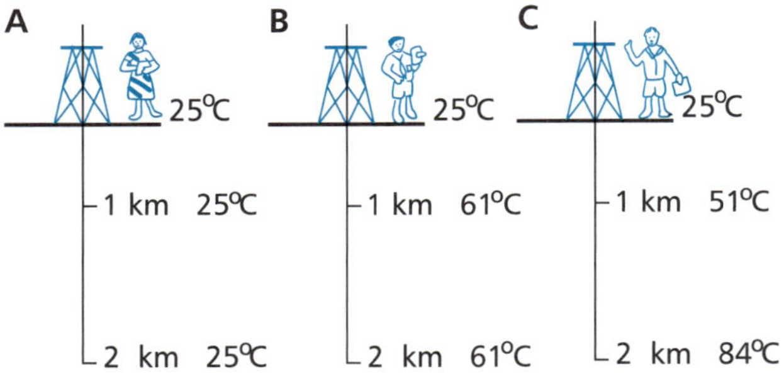

Question 7

a The table below lists a selection of rocks. Fill in the correct rock group — sedimentary, metamorphic or igneous.

Rock	Rock group
marble	
sandstone	
granite	
slate	
basalt	
limestone	

b Match the rock group with its description.

Igneous	rock changed by heat, pressure and/or chemical reaction
Sedimentary	formed from cooled and hardened magma from deep in the Earth
Metamorphic	formed when layers of sediment have cemented together

Question 8

Chlorofluorocarbons (CFCs) were one of the great discoveries of the 1930s. However scientists now believe that they drift to the upper atmosphere and destroy ozone molecules. CFC use in Australia is limited mostly to the following three products.

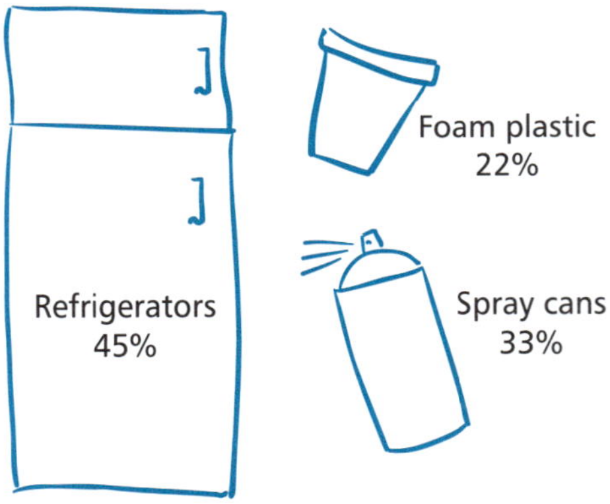

a Draw a histogram to illustrate the data in the drawing.

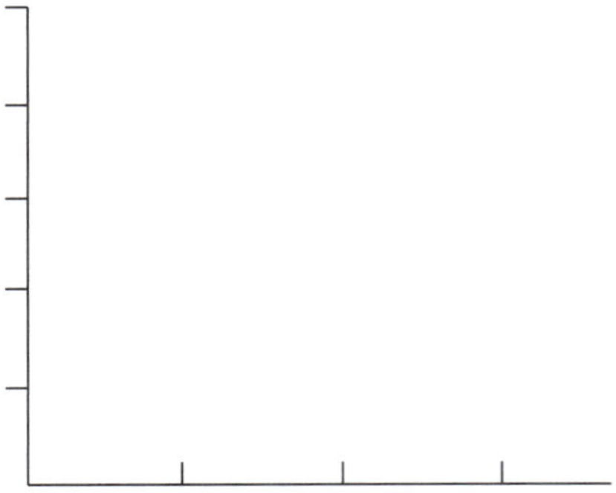

b Why does ozone depletion allow more radiation to reach the Earth's surface?

____

____

c In which major layer of the atmosphere is the ozone layer?

____

Question 9

The heat of the Earth provides the energy that makes the rocks move around and go through a series of changes. The steps in this rock cycle are listed below with the first (1) and the last (8) in the correct place. Put the others in the correct order.

(1) When rocks come into contact with magma they melt.

_____ At the bottom of lakes and seas the bits of rock make sediments.

_____ Surface rocks are broken into bits by the weather.

_____ As sedimentary rocks get buried the temperature and the pressure increase.

_____ As the sediments are covered by new layers, sedimentary rock forms.

_____ Bits of rock are carried down mountainsides by rain and creeks.

_____ Magma solidifies deep underground or at the crust surface to form igneous rocks

(8) Sedimentary rocks slowly change into metamorphic rocks.

Question 10

a Complete the following sentences with the correct words.

You can see only when there is ____________. Things you see either ____________ light into your eyes or give out ____________. Light travels in ____________ ____________, so it cannot go around ____________. When light cannot pass through something, it makes a ____________.

b List three things that give out their own light.

____

____

____

Question 11

We can see things because they reflect light.

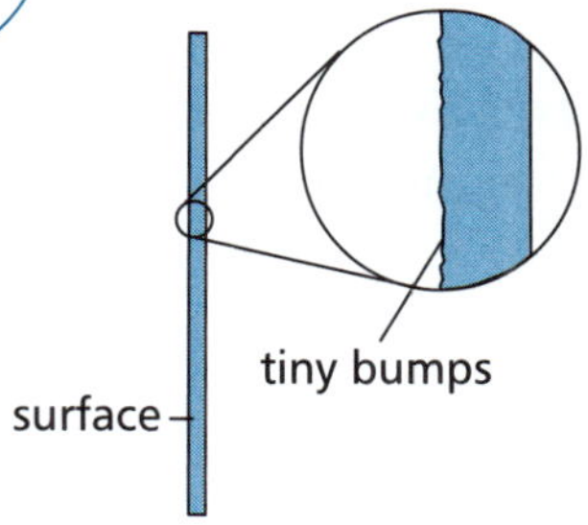

a The surface shown below is bumpy, so it must reflect light in __________ directions.

b Shiny surfaces, like mirrors, reflect rays of light at the ____________ angle as it strikes the mirror.

c Look at the diagram below. If angle *x* is 45 degrees then what is angle *y*?

____

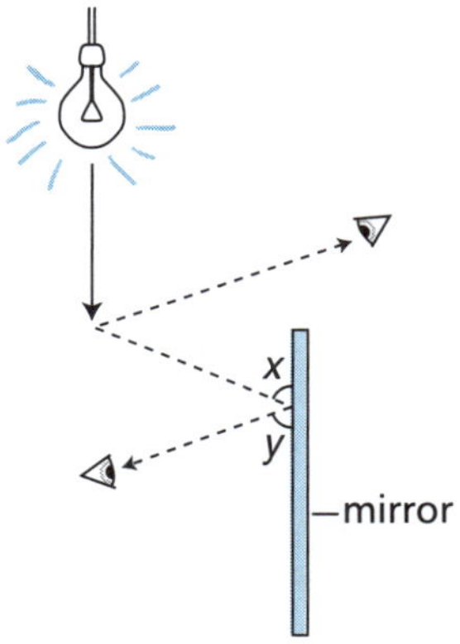

Question 12

White light can be split by rain drops into many colours.

a Write down the colours in order starting with red.

____

____

b Another way to get coloured light from white light is to use a filter. What colour will white light be if it passes through a red filter? Why?

____

____

Question 13

We use light and sound to find out about the world around us.

a We use our ____________ to see the ____________ reflected from things around us and we use our ____________ to hear ____________.

b Why do you hear thunder after you see lightning?

____

____

c Why does light get fainter as it gets further away?

____

____

d People on Earth are able to see an explosion in space but they can't hear it? Why?

____

____

Question 14

a Explain how drums make sounds.

____

____

____

b Look at the diagram and complete the following flow chart.

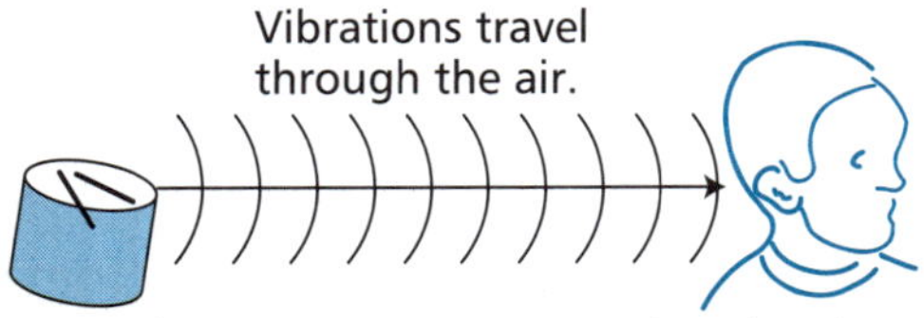

Flow Chart

Drums make ____________.
Vibrations travel through the ____________.
Vibrations enter your ____________.
The ____________ records the vibrations.

Question 15

a Light bends when it passes from water into air. This is called ____________.

b The line between the water and air is called a ____________. A ____________ is a line at right angles (90°) to the boundary.

c When light crosses the boundary from water into air, it is refracted ____________ from the normal.

d When light crosses a boundary at ____________ ____________, it is not refracted.

e Look at the diagram below. Is red light refracted more or less than violet light when it passes into a prism, and then when it passes out of a prism.

____

____

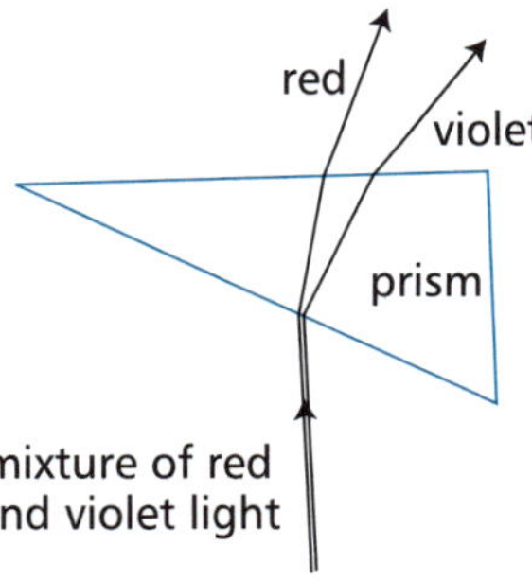

Question 16

Choose the correct word to complete the sentences.

To get vibrations with a higher frequency:

a You can make a string looser/tighter.

b You can make a string shorter/longer.

c You can use a thinner/thicker string.

Question 17

Look at the diagrams below.

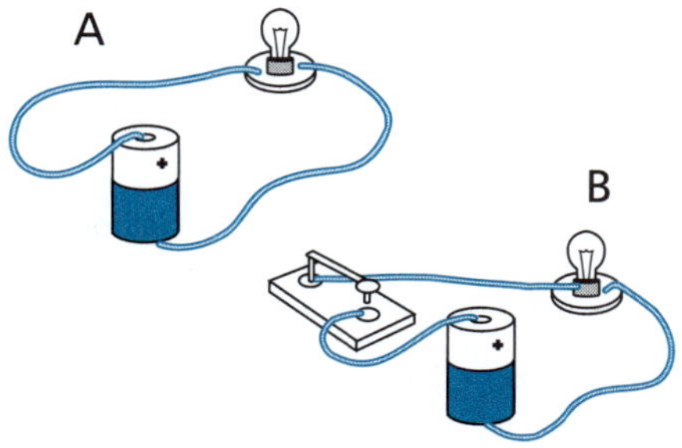

a Is current flowing in A? ____________________

b Is current flowing in B? ____________________

c How could you make current flow in B?

____

d The switch and the bulb are connected in

__________.

Answers

Chapter 1

Science skills page 5

1 a Plant — all yes Soil — all no b no, yes

2 a

Flower	Reproduction
stem	supports the plant
leaf	capture energy and make food
root	take in water and minerals
bud	grow into flowers and leaves

b In order to make their own food, or photosynthesise, plants need light, water and carbon dioxide.

3
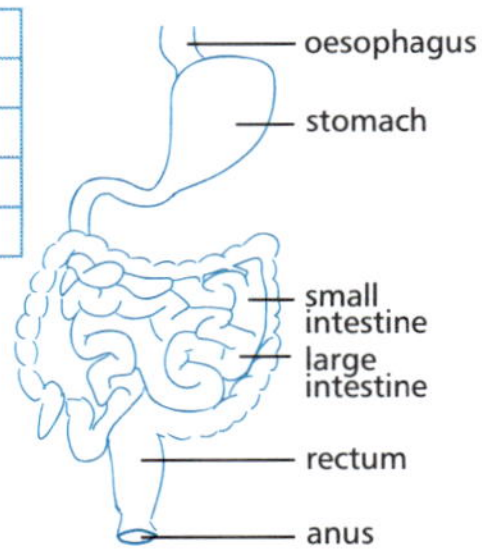

4 touch, taste, smell, sight, hearing

5 Bacteria — absorb food across their cell surface.
Plants — make their own food.
Animals — eat plants and other animals.

Part	Male	Female	Function
anther	✔		makes pollen
filament	✔		supports anther
stigma		✔	pollen lands here
style		✔	pollen grows down the style
ovary		✔	protects ovules
petal		✔	attracts pollinators

7 Vertebrates
- fish — bream
- amphibians — toad
- reptiles — brown snake
- birds — sparrow
- mammals — blue whale

8 Key to plant groups
- makes seeds
  - seeds in cone — conifer (B)
  - seeds in fruit — flowering plant (A)
- makes spores
  - spores in capsule — moss (D)
  - spores under leaves — fern (C)

9 A — whelk C — whelk
B — cerith D — topshell

10 Start Here

Has it got 4 wings → NO → Has it got a long body and long legs → YES → Has it got hairy antennae
- Has it got a long body and long legs → NO **(A)**
- Has it got hairy antennae → NO **(B)**; YES **(C)**

Has it got 4 wings → YES → Are the front wings bigger than the back wings? → NO → Are the wings hairy?
- YES **(D)**
- NO **(E)**

A — house fly C — mosquito E — dragonfly
B — crane fly D — cadis fly

11 1	Fish with a long tube like body	**FISH 1**
	Fish with regular body shape	go to 2
2	Fancy dorsal (top)	**FISH 2**
	Regular dorsal fin	go to 3
3	One dorsal fin	go to 4
	Two dorsal fins	**FISH 3**
4	Spot on its body	**FISH 5**
	No spot on body	**FISH 4**

12 Start Here
- two needles
  - very long needles **(C)**
  - medium needles **(A)**
  - short needles **(D)**
- more than two needles
  - 3 needles **(B)**
  - more than 3 needles **(E)**

13 A — rose C — lily
B — daisy D — carnation

14 Pollen grains land on the stigma.
Pollen tube grows down through the style.
Pollen tube grows down to an ovule.
Male sex cell moves down the pollen tube to the ovule
Fertilisation occurs.

15 a Arachnids have: legs, hard body coating, less than 7 pairs of legs, 4–7 pairs of legs, more than 4 pairs of legs
b Crustaceans also have a hard body coating and 4–7 pairs of legs.
c Arachnids do not have feelers.
d They both have no legs and a thin tube body.
e The annelid has rings on its body.

Revision questions page 9

1 a The parts at the end of the leaves move.
b When an insect crawls into the trap, delicate hairs are irritated and the trap closes.
c Plants cannot move about because their roots hold them in the soil.

2 animals, sense, move, grow, reproduce, waste, energy, water, minerals, carbon dioxide, jobs, organs

3 a organs, systems
b nervous system, digestive system, circulatory system

4 a shark, cat, crab, bee (The larger the animal, the more cells it has.)
b unicellular

5 xylem, water, phloem, food

6 a light, smell, taste, sound, touch
b light and moisture
c in every cell
d carbon dioxide
e oxygen
f urine, sweat, carbon dioxide

7 Airplane or helicopter

9
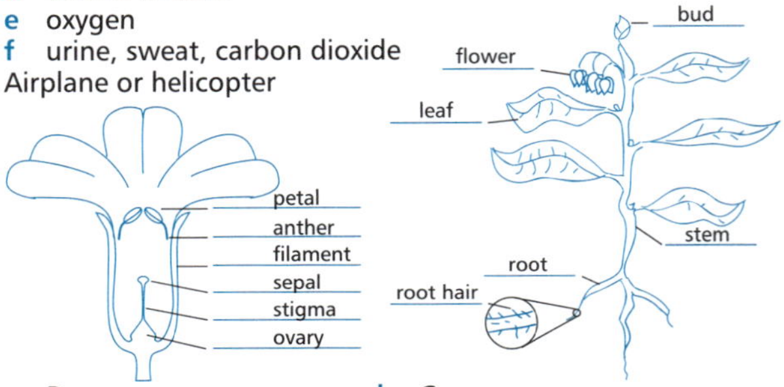

10 a B b C

11 a Start Here
- burrow
  - large tube shape (A)
  - small cone shape (B)
- no burrow
  - hunter (F)
  - lives under bark (C)
  - spins web
    - hairs on legs (D)
    - no hairs on legs (E)

b	1	has a burrow	go to 2
		has no burrow	go to 3
	2	large tube shaped burrow	A
		small cone shaped burrow	B
	3	spins web	go to 4
		does not spin web	go to 5
	4	hairy legs	E
		no hairy legs	D
	5	hunter	F
		lives under bark	C

12 striped — Spencer furry — Rosie

13 fish
mammal
bird
go to 4
has dry skin and scales – reptile
has smooth moist skin – amphibian

14 a A — moves, uses oxygen, gets rid of waste
B — moves, senses, grows, uses oxygen, gets rid of waste, reproduces
C — none
D — moves, uses oxygen, gets rid of waste

b B

15 food and oxygen — waste products

16 They have hair, 4 legs and can run.
The snail has no hair, no legs and cannot run.

17 sawflies = A — grass grubs = C
spring tails = B — soldier flies = D

18 organ, epithelium, tendon, muscle

19 a drinking fluids and eating food
b (i) urine production (ii) sweat and faeces

20 Differences between organisms.
Similarities between organisms.

Chapter 2

Science skills page 17

1
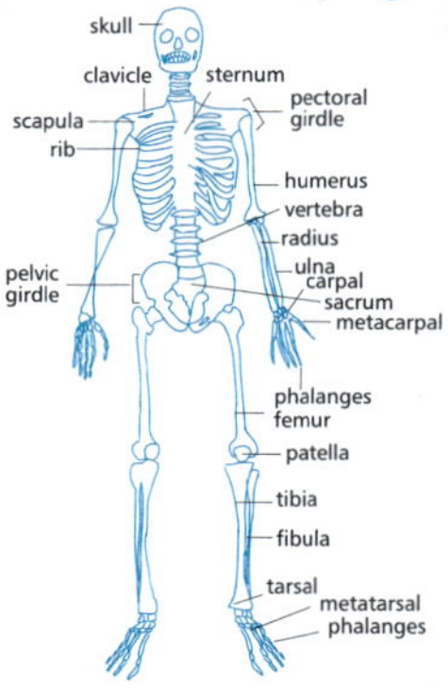

2
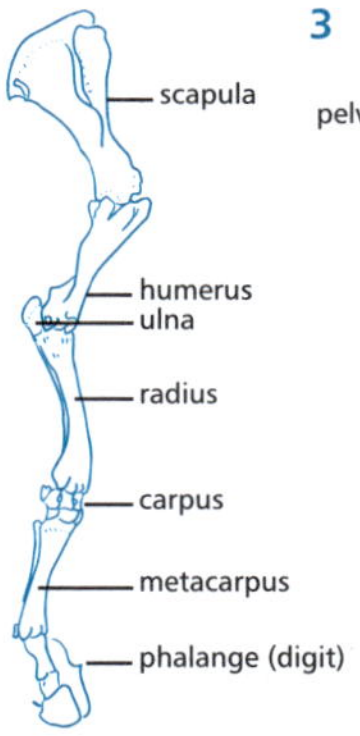

3
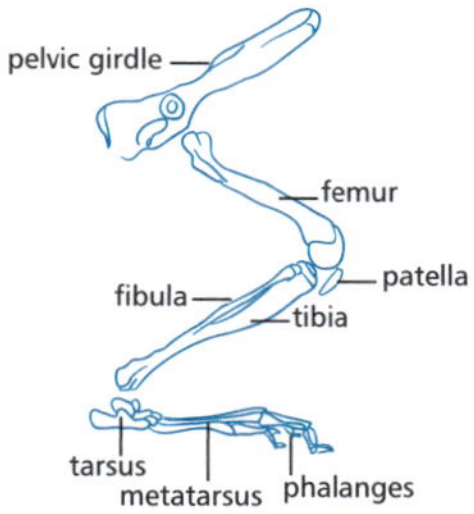

4

Scientific name	Common name
clavicle	collar bone
mandible	lower jaw
sternum	breastbone
pelvis	hip bone
metatarsals	foot bones
tibia	shin bone
patella	knee cap
scapula	shoulder blade

5
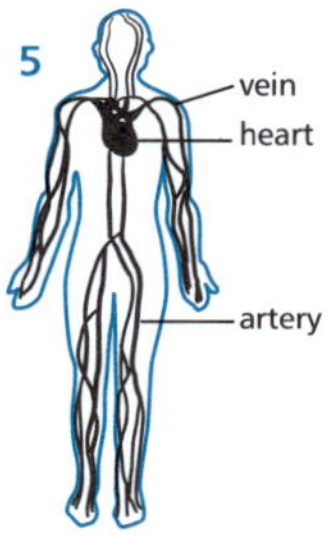

6

Part	Letter	Function
right atrium	A	receives blood from the body
left atrium	C	receives blood from the lungs
right ventricle	B	pumps blood to the lungs
left ventricle	D	pumps blood into the aorta
vena cava	E	brings blood from the body
pulmonary vein	H	brings blood back from the lungs
aorta	G	delivers blood to the body
pulmonary artery	F	takes blood to the lungs

7 oxygen molecule in the lungs
→ pulmonary vein carries oxygen to the left atrium of the heart
→ left ventricle pumps oxygen into the aorta
→ aorta carries the oxygen into the capillaries of the muscle
→ capillaries deliver the oxygen to the muscle cells

8
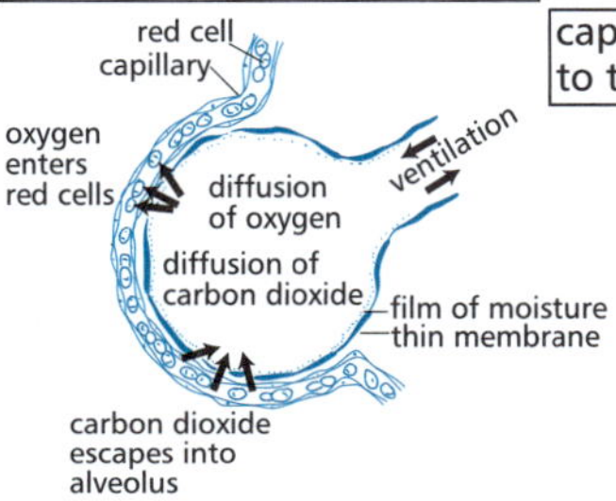

9
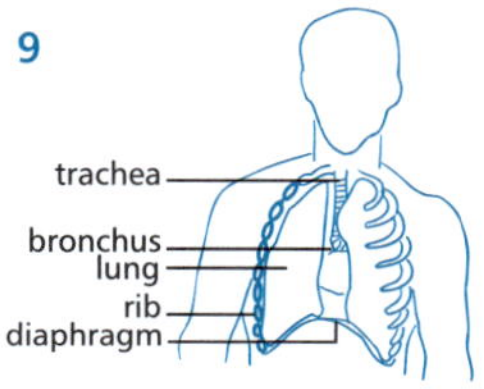

10 Light the marshmallow. If it burns then it supplies energy.

11

Part	Letter	Function
small intestine	E	digestion and absorption of food
oesophagus	H	transport food from mouth to stomach
liver	B	store food and neutralises acid
stomach	G	break up of food, protein digestion
pancreas	F	produces hormones & enzymes
anus	C	expulsion of waste
salivary glands	A	produce saliva for lubrication & starch digestion
large intestine	D	water absorption & compaction

12 a proteins, glucose b urinary bladder c urine

13 a
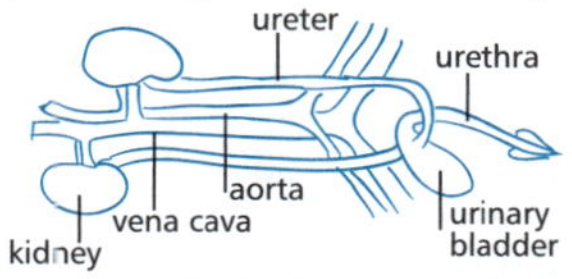

b Because the kidneys are the organs that filter the waste from all the body's blood.

14

System	Organ	Function
respiratory	lungs	releasing energy from food
digestive	stomach	breakdown of food
excretory	kidneys	getting rid of waste
skeletal	bones	framework that supports the body
circulatory	veins & arteries	maintains a constant flow of blood

15 cells — Collections of basic elements.
tissue — Group of similar cells specialised for one simple job.
organ — Collection of different cells grouped together to perform a particular function.
organ system — Collection of organs performing a main body function.

Revision questions page 21

1 a urine (kidneys) b sweat (skin) c carbon dioxide (lungs)

2 a transport oxygen in the haemoglobin b fight infection
c assist clotting of blood d fluid part of the blood

3 diaphragm, increases, decreases, diaphragm, decreases, increases

4 a The movement of oxygen into the blood and carbon dioxide out.
b Gas exchange occurs in the alveoli of the lungs.
c To supply oxygen for the combustion of fuel to supply energy for the body.

5 Respiration — chemical reaction to breakdown food
Nutrition — obtaining raw materials to sustain life
Excretion — expulsion of waste
Circulation — blood flow
Skeleton — provides support and an anchor for muscles

6 A synovial joint — B slightly moveable joint
C fixed joint — D synovial joint

7 a A non-collapsible tube connecting the transporting air to the bronchi.
b Air carrying tubes which divide into ever decreasing sizes until the smallest — bronchioles.
c The smallest subdivisions of the air passages — site of gas exchange.
d The main respiratory organ.

8 a carpal, metacarpal, phlanges — b humerus & scapula
c femur — d ribs & vertebrae
e tibia & fibula — f radius & ulna
g pelvic girdle

9 a man — b elephant
c dog — d horse

10 a To rid the body of unwanted waste.
b Nitrogenous products (like urea & uric acid); carbon dioxide
c To remove the waste materials and help maintain water balance in the animal.
d The kidney of the desert animal must be able to reabsorb nearly all the water that enters the kidney. The kidney of the rainforest animal will produce large quantities of dilute urine.

11 a A — circulatory, B — excretory, C — digestive, D — respiratory

b A heart, arteries, veins
B kidneys, bladder
C stomach, small and large intestines
D lungs

12 a vein b capillary c artery

13 oxygen, energy, respiration, carbon dioxide, lungs

14 a Because the muscles need more oxygen and more nutrients when they are exercising.
b pulse
c It is not pumped to the extremities as the body tries to conserve heat for the main part of the body where all the organs are.

15 a Because there are so many cells in multicellular organisms, most are a long way from the external environment.
b i Carry oxygen and carbon dioxide to and from cells.
ii Remove metabolic waste from body cells.
iii Transport organic materials (nutrients) to body cells.
iv Transfer heat around the body.

16 a right side
b to become oxygenated
c arteries carry deoxygenated blood and the veins carry oxygenated blood
d arteries
e No — veins do not have a pulse.

17 a This large surface area with a very thin but extensive blood supply ensures that there is rapid and efficient exchange of gases in the lungs.
b i Moist — to allow gases to be in solution
ii Thin — to let the gases through
iii Large surface area — for efficient and fast exchange.

18 peristalsis

19 a To breakdown complex molecules and to absorb the products.
b Because food for humans is made of organic molecules that are different from human organic molecules and must be broken down to the basic units (sugar, lipids, amino acids).
c Immediate food storage, some digestion and absorption.
d Controls movement of food into the small intestine.

20 Digestion, teeth, stomach, enzymes, Amylase, pepsin

Chapter 3

Science skills page 29

1 Telophase — B
Metaphase — C
Early prophase — A

2 A = second division
B = first division

3

	Chromosomes separate	Chromatids separate
Mitosis		*
Meiosis (Division 1)	*	
Meiosis (Division II)		*

4 a 45 in every 10 000 people will die.
b 10 in every 10 000 people die.

5

Health Problem	Cause
obesity	too much food
poor eyesight	vitamin A
anaemia	not enough iron
weak bones	vitamin D
anorexia nervosa	too little food

6 Process A = Meiosis
Process B = Mitosis

7

Disease	Cause
sickle-cell anaemia	inherited condition
diabetes	too much sugar
deafness	noise pollution
obesity	too much food
lung cancer	smoking

8 a 4 hours b 8 hours c 2 grams

9 a cell A b zygote

10

Girls	Boys
body hair grows	body hair grows
ovaries make ovum	testes make sperm
breasts develop	penis grows bigger
monthly periods begin	voice deepens

11

Weeks since fertilisation	Size (cm)	Name given to the stage of development
4	1	embryo
8	3	foetus
12	12	foetus
38	52	baby

12

Characteristic	Somatic cells	Sex cells
they make up almost all of the body cells	*	
they divide by meiosis		*
23 chromosomes		*
diploid	*	
not involved in reproduction	*	
pass on genes		*

Revision questions page 31

1 meiosis, chromosomes, same, somatic

2 a true b true
c false d true

3 a meiosis b fertilisation

4 heart, oxygen, waste, carbon dioxide, body weight and strengthens bones

5 a It stops skin cancer.
b No.
c Cells growing out of control.

6

Disease	Example
inherited	cystic fibrosis
nutritional	anaemia
physiological	diabetes
environmental	cancer

3

7 prophase, metaphase, anaphase, telophase, interphase

8 crossing over, genes, genes, parent

9 1. physical exercise
2. good eating habits
3. no smoking
3. reduce stress
5. no alcohol
6. no drugs

10

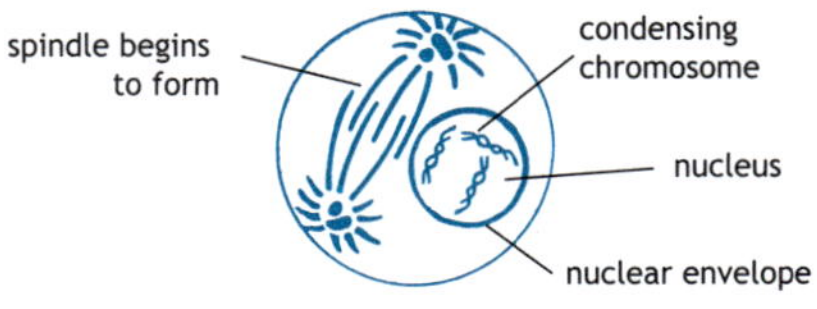

11 A = Anaphase
B = Interphase

12

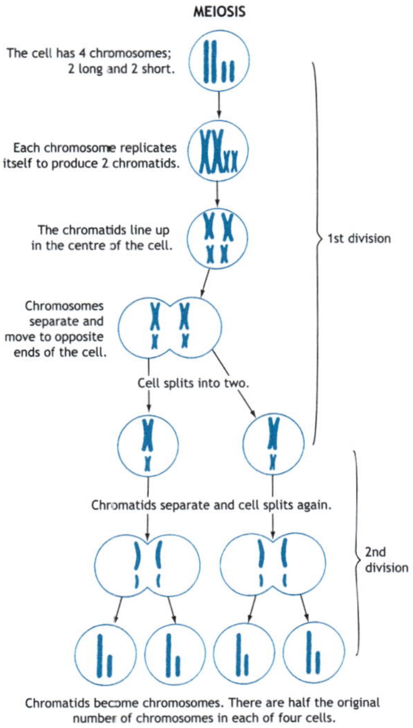

13 a Mitosis and meiosis.
b Mitosis.
c They remain the same.
d The nucleus.
e

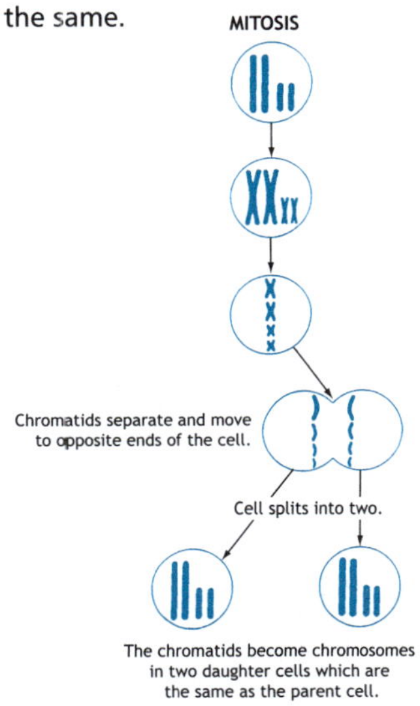

14 a They keep on multiplying long after the tissue has been produced or renewed.
b A tumour is a group of out-of-control cells that form a lump.
c A substance that causes cells to become cancerous.
d Asbestos and smoking are both carcinogens.

15 a, b and d.

16 b

17 b

18 c

19 c

20 diabetes, insulin, obesity, sugar, exercise

Chapter 4

Science skills page 39

1

Property	Solid	Liquid	Gas
Takes up space	yes	yes	yes
Has weight	yes	yes	yes
Fixed volume	yes	yes	no
Fixed shape	yes	no	no
Strength	yes	some	no
Crystals	some	no	no
Can be compressed	no	no	yes
Density	high	medium	low

2

Object	Solid	Liquid	Gas
Glass		✔	
Shaving foam		✔	✔
Soft drink		✔	✔
Perfume		✔	
Syrup		✔	
Chalk	✔		
Cotton wool	✔		
Air			✔
Rubber band	✔		

3 a yes
b Because there are spaces between the gas molecules and these can be compressed.
c Yes, because the moving molecules will push it back out again.
d No, because milk is denser than air, and there are practically no spaces between the molecules.
e No, because there are no spaces.
f The number of particles/mL and the density will increase.

4 a wood = (1.74 divided by 2) = 1.76 g/cm^3
cheese = (1.74 divided by 2) = 0.87 g/cm^3
cork = (1.76 divided by 8) = 0.22 g/cm^3
b cork, cheese, wood

5 a 1.3 g/litre at normal temperature and pressure.
b Water is 769 times denser than air.
(Divide the mass of 1 litre of water by the mass of 1 litre of air — 1000 divided by 1.3)
c Volume of the garden shed = 12 m^3.
d 15 600 g or 15.6 kg
(Multiply the volume in litres by the mass of 1 litre of air — 12000 × 1.3)
e Hydrogen is the lightest gas but it burns so it is not the safest gas. Helium is also lighter than air and it is safer.
f Carbon dioxide is heaviest so gravity pulls on it more than hydrogen or helium.

6 Solid chocolate — Diagram b
Melted chocolate — Diagram c

7 a

	Part of graph		
	A to B	B to C	C to D
Temperature	increasing slowly	increasing faster	constant not changing
Particles	vibrating a little quicker	vibrating faster	vibrating very fast
State	solid	liquid	gas

b Take the energy away — put in the freezer.

8 a D, A, C, B
b The more molecules in the same space the higher the pressure.

9

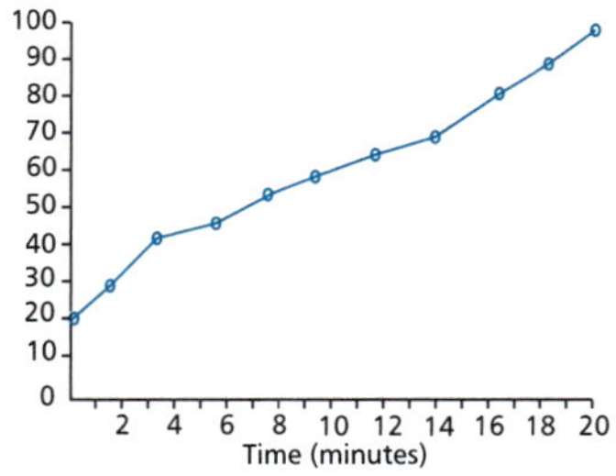

10 a diamond
b cardboard
c All the same mass, as the volume of each is the same.
d Diamond — it has the most molecules in a given volume.
e cardboard
f bone, clay, diamond, slate
g Because ice is less dense than water.
h bone

11 a solid
b gas
c liquid

12 a The balloon will inflate because the air molecules will expand and take up more space.
b The balloon will deflate because the air molecules will contract and take up less space.
c Hot air balloons

13 a Yes
b Fill up 5 test tubes with different liquids (e.g. cooking oil, water, honey, milk, fruit juice) to the same level.
Drop a paper clip into the first tube and time how long it takes to get to the bottom. Repeat for the other tubes.

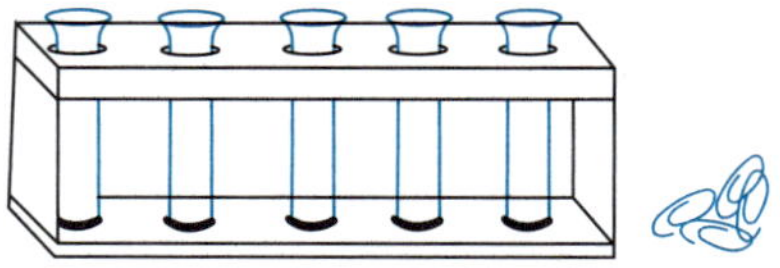

The paper clip will get through the water fastest because there are more spaces (holes) between the molecules in water than in denser liquids like honey.

Revision questions page 42

1 a solid — the molecules are packed closely together, which increases the density
b gas

2 diffusion

3 a sublimation
b one — solid to a gas
c It would change straight to a gas and not become water first.

4 A — condensing
B — melting
C — boiling
D — freezing
E — evaporation

5 Because air has lots of spaces between the molecules and we can push between them easily, but a fence is a solid with no spaces so we can't push between the molecules.

6 It will spread out (diffuse) everywhere.

7 a B
b B

8 a They both weigh the same.
b butter
c The kilogram of butter weighs more. The butter container full of rice bubbles weighs less as there are less molecules in the same space.
d A light rock has less molecules and therefore takes less energy to move.

9 a Fat people as they are denser.
b Boys as they have less fatty tissue than girls.
c Because they can change their density by adjusting how much air and water they are carrying.
d Because it has gone rotten and now contains smelly gases which are less dense.
e As the water becomes warmer, it will rise above the Plimsoll line because it is denser than cold water. This may cause the ship to sink.
f Because it has salt molecules in it as well.
g Because the moisture heats up and the molecules take up more space and can no longer fit in the kernel so it explodes.

10 Nothing will happen because the adhesion (between the mercury and glass molecules) is less than the cohesion between the mercury and the glass molecules.

11 a In both a liquid and a gas the particles can move around in all directions.
b Particles in a gas are much further apart than (they also move around much faster and can move into all the spaces that are available).

12 faster, vibrate, liquid, melted

13 a When a liquid absorbs enough energy (from the sun) it changes into a gas. The water changes to water vapour and spins around in the air — evaporation.
b No
c Clothing will dry faster as there is less water to evaporate.
d No

14 C — it has a larger surface area so particles moving fast enough can escape from the liquid at a more rapid rate.

15 Because cold air cannot hold as much water vapour as warm air, it condenses as the air is cooled and drips out.

16 The water on the surface of our body evaporates into the air. The energy that the particles need to escape is transferred from our body so we feel cool.

17 An iron nail expands when it is heated because the vibrating particles are colliding with each other with greater force, and take up more space.

18 a Surface tension will pull the thread back towards the remaining film.
b There is a surface force or surface tension on the surface of a liquid due to forces attracting each surface molecule to its neighbour.

19 a add
b add
c subtract
d subtract
e add

20 a On hot days the concrete sections expand and the squashy bitumen stops the concrete cracking as it expands.
b If the wires contract in cold weather they get tighter but do not snap.

Test A page 47

1

Vertebrate type	Example
fish	bream
amphibian	frog
mammal	goat
reptiles	crocodile
birds	willy wagtail

2

Group	Description
crustaceans	2 pairs of antennae, five or more pairs of legs
insects	3 pairs of legs, antennae
spiders	4 pairs of legs, no antennae
many legs	long body divided into segments, legs on every segment

3 a protists, monerans, fungi, plants, animals, invertebrates, vertebrates
b sea anemone, heart worm, earthworm, oyster

4 A — guppy
B — barb
C — catfish
D — swordtail
E — molly

5 a

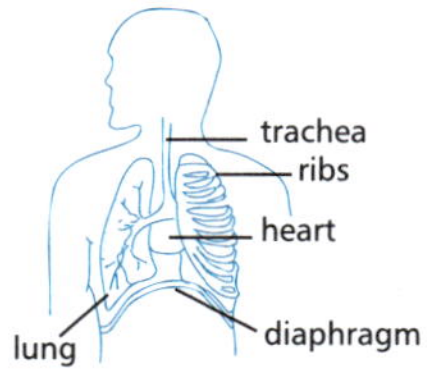

b alveoli, Oxygen, capillary, Carbon dioxide, alveoli

6 A synovial (ball & socket) joint
B slightly moveable
C synovial (hinge) joint

7 Your heart pumps blood to your body cells

Body cells swap carbon dioxide for oxygen.

Blood from the body cells returns to the right side of your heart.

Your heart pumps blood to your lungs.

Blood stores oxygen and gets rid of carbon dioxide.

Blood leaves your lungs and travels back to the left side of your heart in veins.

8

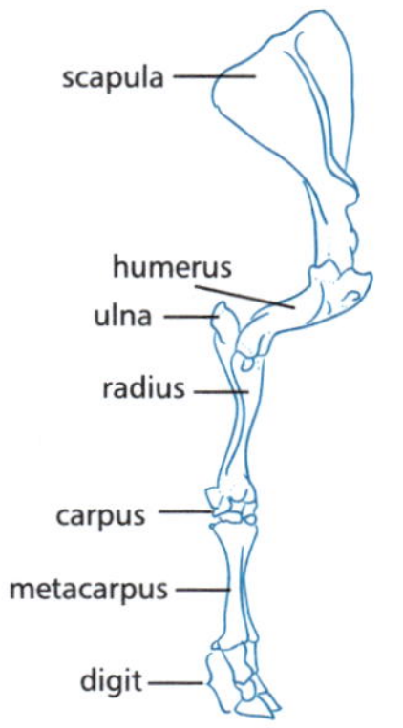

9 a and b

Prophase — Paired chromatids untangle, shorten and thicken and the spindle forms.

Metaphase — Nuclear membrane disappears, chromosomes line up in the centre and the chromatids move to the opposite ends of the spindle.

Anaphase — Chromatids complete movement to the ends of the spindle.

Telophase — Cell membrane begins to constrict and two daughter cells are formed.

10 a metaphase b anaphase
c telophase d prophase

11 a, c and d

12 cell division, testes, ovaries, halved, reinstated, fertilisation

13 a b

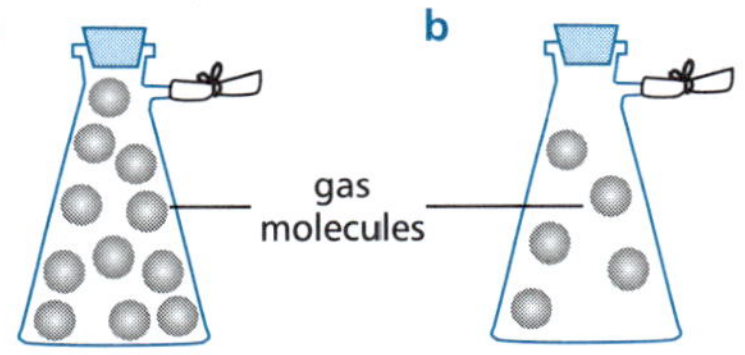

14 a (ii)
b Molecules in B and C are packed closer together, they are denser liquids and so they weigh more.

15 a evaporation e melting
b condensation f condensation
c melting g condensation
d freezing h evaporation

16 Solids —The particles cannot move about but they vibrate. The particles are very close together.
Liquids — The particles move about slowly. The particles are quite close together.
Gas — The particles are far apart. The particles move about very fast.

17 a, d, e, f, g, h

18 a energy b energy c liquid

19 a

Substance	Density (g/cm^3)	Mass (g)	Volume (cm^3)
A	2	4	2
B	8	32	4
C	2	1	0.5
D	0.5	2	4

b B d A and C
c C e D

20 a dry day — as the air can hold more water vapour
b hot day — as there is more energy (heat) in the air to cause evaporation
c windy day — as there is more air movement
d spread them out to increase the surface area for evaporation

21 Fill the four jars to the same level with the liquids. Measure the time it takes for a marble to reach the bottom of each jar.

Chapter 5

Science skills page 56

1 a H = hydrogen, S = sulphur, O = oxygen
b sulphuric acid
c The number of atoms of each element.
d The atomic composition of the substance.
e atom

2 sodium and chlorine → sodium chloride

3 a iron + oxygen → iron oxide
b iron and oxygen
c iron oxide
d magnesium + oxygen → magnesium oxide
e magnesium oxide

4 a salt (sodium chloride) and water
b hydrochloric acid + sodium hydroxide → sodium chloride + water
c $HCl + NaOH \rightarrow NaCl + H_2O$
d neutralisation reaction
e Because the H^+ ions are neutralised by the OH^- ions in the sodium hydroxide and 'tied up' as neutral water.

5 a no b zinc chloride c no
d hydrogen e $Zn + 2HCl \rightarrow ZnCl_2 + H_2$

6

Substance	What atoms is it made of?	Element or compound?
copper	copper	element
water	hydrogen and oxygen	compound
hydrogen	hydrogen	element
copper oxide	copper and oxygen	compound
carbon dioxide	carbon and oxygen	compound

7 B

8 a B
b corrosion
c oxygen
d To stop air getting into the boiled water.
e moisture and oxygen

9

Substance	Chemical formula	Types of atoms (no.)	Atoms in total
hydrogen gas	H_2	1	2
water	H_2O	2	3
sulphur	S_8	1	8
methane	CH_4	2	5
vinegar	CH_3CO_2H	3	8
sulphuric acid	H_2SO_4	3	7
alcohol	CH_3CH_2OH	4	9
carbon dioxide	CO_2	2	3
nitrogen gas	N_2	1	2

10 a $2HCl + Na_2CO_3 \rightarrow 2NaCl + H_2O + CO_2$
b positive, negative, attract
c $Na_2^+CO_3^{2-}$
Na^+Cl^-
$Mg_3^{2+}N_2^{3-}$
$K_3^+N^{3-}$
(F_2O_3 is not ionic)

Revision questions page 58

1 a hydrogen and oxygen
b 2 hydrogen atoms and 1 oxygen atom.

2 equation, left, right, react, new

3 a NaCl
b Sodium has two full electron shells but its outer orbit has only one electron. Chlorine has seven electrons in its outer orbit, so it needs one more to make it stable. These elements share an electron which makes their outer shells full.

c $Na^+ Cl^-$
d chemical bond

4 Precipitation – The formation of an insoluble solid when a reaction takes place in a solution.
Combustion – The chemical change in which oxygen reacts with another substance.
Neutralisation – A reaction in which an acid and an alkali react in an aqueous solution to form a salt and water.

5 a NaOH and H_2
b Na and H_2O

6

Ionic	Molecular
neutral	neutral
made up of cations and anions	made up of molecules
mostly crystalline	mostly gases and liquids
made up of metallic and non-metallic elements	made up of non-metallic elements
eg: **sodium chloride**	eg: **water**

7 charges, Mg^{2+}, negative, Cl^-, two, $MgCl_2$

8 In a chemical reaction no atoms are destroyed nor are any created. They are just rearranged.

9 a word equation
b molecular equation
c ionic equation

10 d

11 a hydrogen + oxygen → water
b carbon + oxygen → carbon dioxide
c copper chloride → copper + chlorine

12 a copper + oxygen → copper oxide
b magnesium + oxygen → magnesium oxide
c copper sulphate + sodium hydroxide → copper hydroxide + sodium sulphate
d zinc + sulphuric acid → zinc sulphate + hydrogen
e hydrochloric acid + sodium hydroxide → sodium chloride + water

13 a combustion reaction
b combustion reaction
c precipitation reaction
d neutralisation reaction
e neutralisation reaction

14 b

15

Example	Reaction
respiration	combustion
rusting	corrosion
relieve indigestion	neutralisation
burning fossil fuels	combustion
managing soil pH	neutralisation

16 a compound
b precipitate
c chemical reaction
d oxide
e chemical bond

Chapter 6

Science skills page 65

1 a

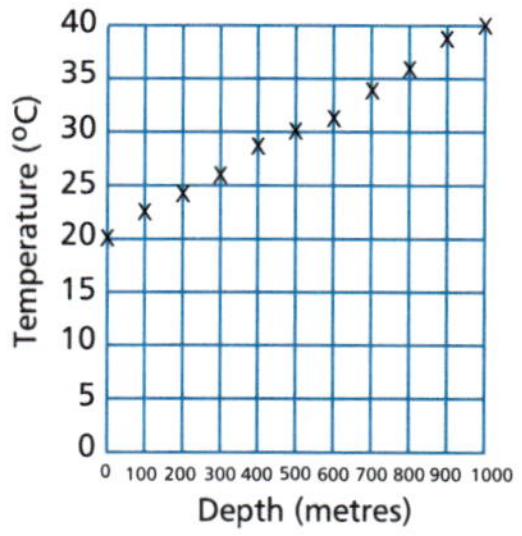

b 20 °C
c 40 °C
d 1000 metres
e The supervisor is getting closer to the centre of the Earth.

2 1 basalt
2 sedimentary rock
3 metamorphic rock
4 magma
5 granite

3 a Streams and rivers carry bits of rock to the sea.
b The bits of rock (called sediment) settle at the bottom of the sea.
c Older layers of sediment get squashed by the newer layers.

4 a igneous — granite
sedimentary — chalk, sandstone, limestone
metamorphic — marble, slate
b Chalk, limestone and marble are all made from calcium carbonate.
c sandstone
d slate

5 The surface of the rock should start to crack and break off in little bits.

6 marble, coral, chalk, limestone

7 a A
b The jar represents the Earth's atmosphere.

8

physical		chemical		biological
water	wind	rusting	acid rain	plants
ice				animals

9

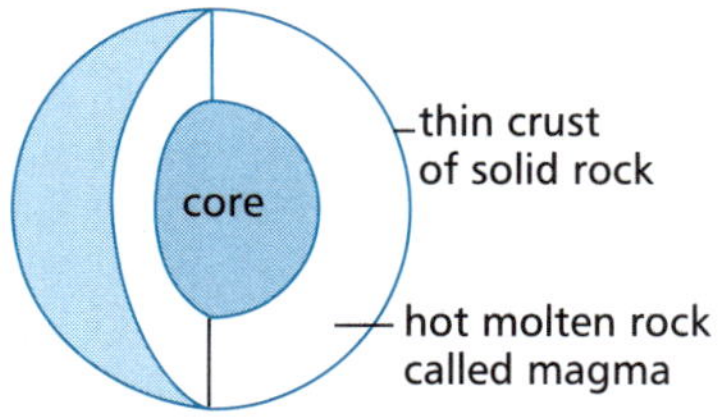

Revision questions page 68

1 b

2 d

3 a granite
b basalt
c igneous rock

4 a marble
b slate
d metamorphic rocks

5 weathering, cracks, expands, break

6 a Carbon dioxide combines with water to form the weak acid carbonic acid.
b The rocks wear away by the chemical reaction between the calcium carbonate in these rocks and the acid.
c chemical

7 b

8 a oxygen, nitrogen b nitrogen c carbon dioxide

9 a fresh water, air, ultra violet protection, keeps us warm
b skin cancer, eye damage, sunburn causing skin ageing

10 a A blanket of gases that surround our planet.
b troposphere
c Incoming solar radiation is absorbed by the Earth. Some is bounced back and it is absorbed by water vapour, carbon dioxide, etc which increases the global temperature.
d

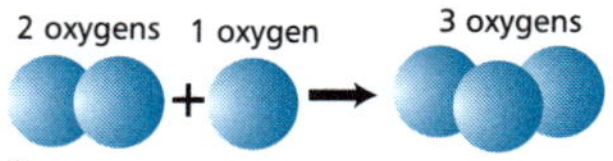

e B

11

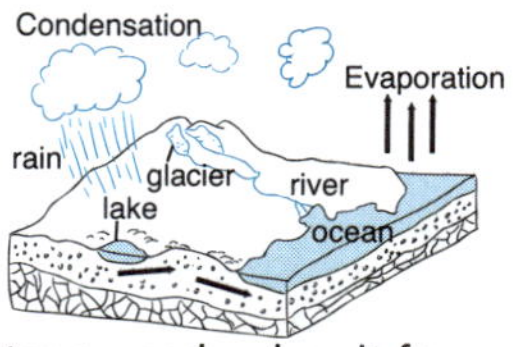

12 Water expands when it freezes. This expansion in the crevices of rocks helps to break the rock into smaller and smaller bits and then into soil.

13 a sedimentary and metamorphic
b igneous and metamorphic
c igneous

d heat and pressure
e yes
14 d
15 carbon dioxide, fossil fuels, temperature
16 a The atmosphere is made of air (includes many gases, dust, pollution).
b The crust is made of solid rock (igneous, sedimentary and metamorphic).
c The magma is molten rock.
d The core is made of iron and nickel.
17 a false b true c true
d true e false
18 neap tides
19 air pressure sea level
20 a atmosphere
b Weathering is the breakdown of rock when it is exposed to the sun, the wind and the rain.
c Erosion is the movement of rock fragments and soil by glaciers, wind, waves and running water.
d The weather

Chapter 7

Science skills page 77

1 a Chemical c Kinetic
b Heat d Kinetic

2

Good heat conductors	Poor heat conductors
copper	wood
steel	glass
iron	plastic
water	wool
brass	rubber

3

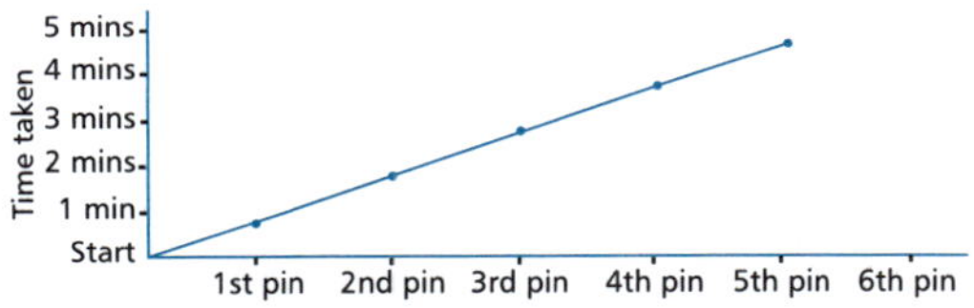

The result is a straight line so heat must flow along the metal rod at an even rate.

4

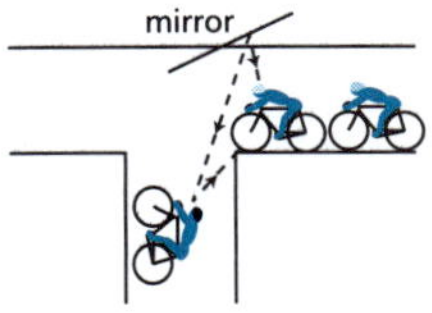

5 a B = 45, X = 45, Y = 45, Z = 90 b 90, 45, 45

6 a

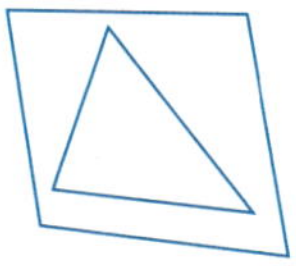

b Light travels in straight line so the shadow is much larger than the wooden triangle.

7 a

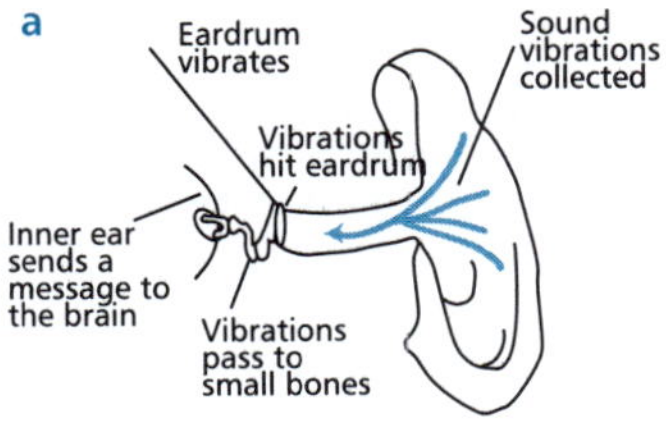

b 1, 4, 2, 3, 5

8

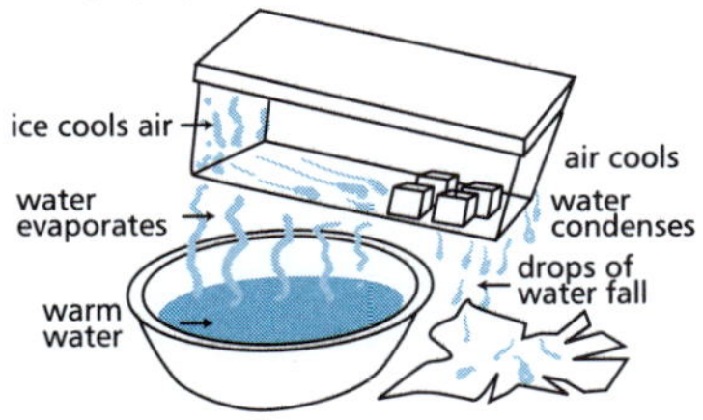

9 a chemical → kinetic → light & heat
b chemical → light & heat
10 a shiny can b dull black can

Revision questions page 79

1 a thermometer
b energy, hotter, colder
c molecules
d Conduction
e reflect
2 Twisting the band builds potential energy. When it is released the stored energy in the band makes the band unwind causing the can to roll. Potential energy becomes kinetic energy. When the potential energy is all used up the can stops.
3 a Yellow light is reflected.
b Green light is reflected.
4 white
5 a It is not as light on a cloudy day because clouds reflect some of the light from the sun back into space. They also absorb some light.
b You don't get strong, sharp shadows on a cloudy day.
6 light, absorb, straight lines, corners, shadow
7 a You see lightning before you hear thunder because light travels through the air a lot faster than sound does.
b In 4 seconds light travels 1320 metres (4 × 330)
8 a and b

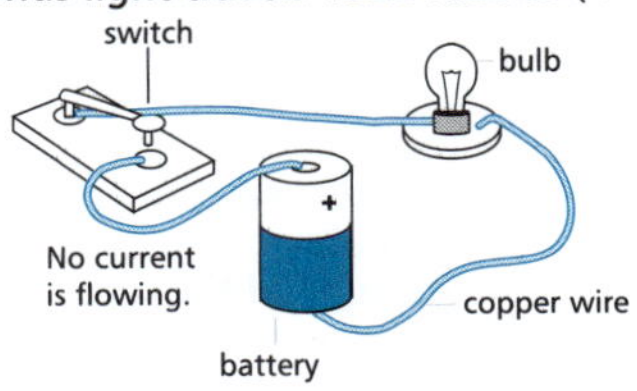

c The bulb lights up. d no e insulator
f open a gap in the circuit (break the circuit)
9 We burn fuels to make things hotter.
We burn fuels to produce light.
We burn fuels to move things around.
10 a From the food you eat.
b Carbohydrates (sugar, bread, pasta, rice) and fats (butter, chocolate, icecream).
c chemical energy
11 A — potential C — potential
B — kinetic D — kinetic
12 kinetic (25%) + heat (65%) + sound (10%) = 100%
13 a Lungs are using chemical energy from food to force air into the balloon.
b Energy is stored in the springiness of squashed air inside the elastic balloon.
c Elastic energy is transferred to kinetic energy (blast of moving air) when the balloon bursts.
14 a potential b electrical c heat d kinetic
15 torch = electrical to light
battery = chemical to electrical
lift going up = electrical to kinetic
solar cell = light to electrical
TV = electrical to light and sound
bonfire = chemical to heat and light
nuclear power plant = nuclear to electrical
16 a The paper dots start moving around in water currents.
b convection
17 a all potential and no kinetic
b half potential and half kinetic
c no potential and all kinetic
18 a violin, guitar, piano c saxophone, oboe, clarinet
b drums, symbols, xylophone
19 a A (smallest amount of air)
b Because different amounts of air vibrate differently making different sounds.

20

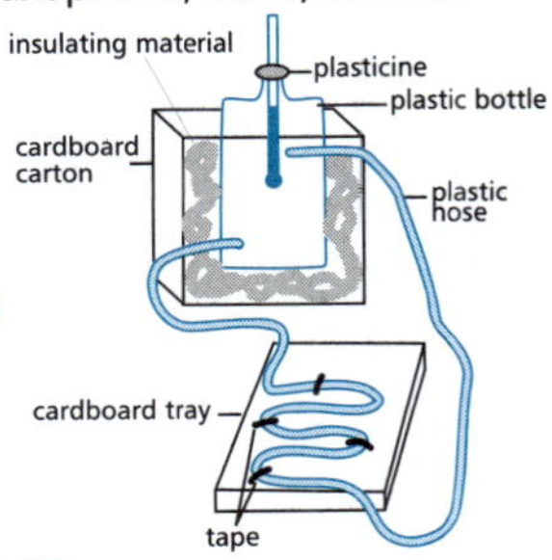

Chapter 8

Science skills page 87

1 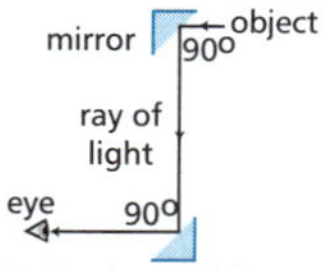

Reflection of light in the periscope

2 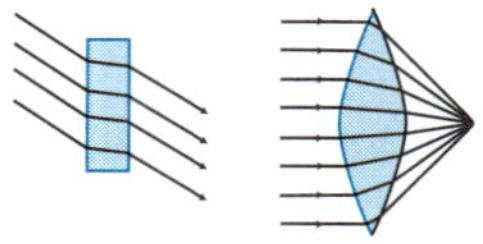

3 d

4 The man is standing with the edge of a mirror against the centre of his chest. The image of half his body is in the mirror.

5 a To focus the signals. b Concave
 c In a high uninterrupted area.
 d A bigger signal as more signals would be focused at the signal detector.

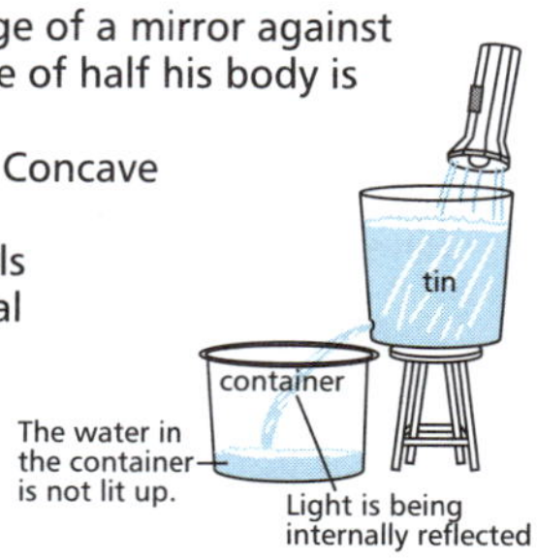

The light is bouncing off the sides of the stream of water. It is being internally reflected at the air/water interface. Fibre optics work this way.

7 a Refraction
 b Light travels faster in the air than in water (air is less dense), and when it comes out through the surface of the water, it changes direction.
 c One wheel will hit the edge of the road and the car will be refracted – ie it will change direction and slow down because the mud is less dense than air.

8 a 45/45, 23/23, 87/87 b reflects, same

9 a A b They have a big amplitude.

10 a 20 Hz c Above 2000 Hz.
 b It falls to about 10 000 Hz.

11 B – it has a high frequency and a small wavelength.

12 A shorter pipe produces a higher pitch (frequency).

13 a 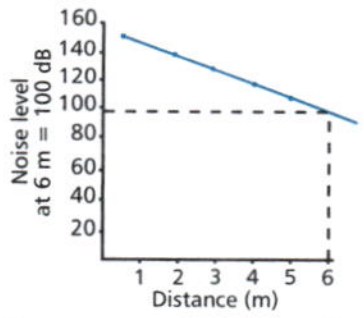

 b No — safe level is 80–90 dB.

14 a A b 4000 × 0.5 × 1 = 2000 m

15 For the bulb to light up, electrons must flow in a complete circuit from the negative terminal to the positive terminal of a battery: A, B, C, H, I

16 C

17 Circuit A — connected in series.
 Circuit B — connected in parallel.
 Circuit C — connected in parallel.

18 In circuit A only bulb 1 comes on.
 In circuit B both bulbs come on.

19 a The filament is broken.
 b Bulb B is connected in series with bulb A.
 Since bulb A has a broken filament so electrons (current) cannot flow through — the circuit is broken.

20 repel, attract, repel

Revision questions page 91

1 a shallower, bent, refracted, normal, air
 b away, towards, right angles

2 a These have reflectors at right angles inside them so they reflect the light from the headlights back towards the cars.
 b Because the plastic is red and acts like a filter, only letting the reflected red light through.

3 a Mirror B
 b The mirror is at the wrong angle to reflect light into the driver's eyes.
 c Adjust the angle of the mirror.

4 a 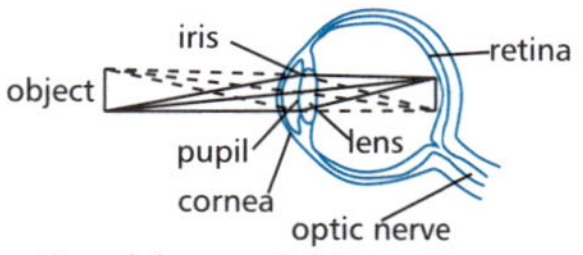

 b retina (e), pupil (c), optic nerve (d), lens (b), iris (a)

5 No, because light travels in straight lines.

6 A — Because light is reflected back at the same angle it hits the surface.

7 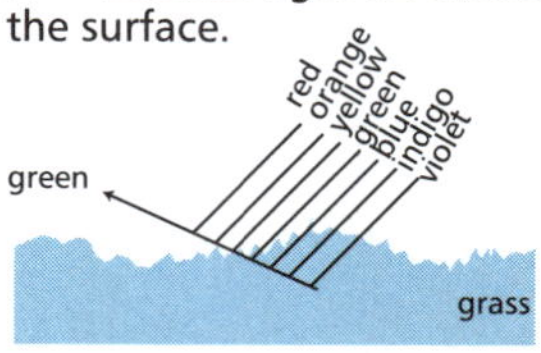

8 Yes, the bell will ring but he cannot hear it, because sound cannot travel through space (vacuum) – it needs particles.

9 The tuning fork sets the molecules vibrating, the ear collects these vibrations, making the eardrum move. The small bones transfer these vibrations to the inner ear and on to the brain.

10 a Violin, as it is a smaller instrument and vibrations have a higher frequency.
 b At the left-hand end.
 c Strings are thicker and longer at the left-hand end.
 d Tapped

11 a They damage the inside of the ear and messages are not passed onto the nerves very well.
 b Wear ear plugs or ear protectors.

12 Sound travels slower than light.

13 a light d light
 b sound e sound and light
 c sound

14 series, parallel, series, series, parallel

15 The bulbs are connected in series. If one light bulb is broken or removed the circuit is no longer complete and the whole line goes out.

16 a A has an insulator in the circuit so electrons cannot flow.
 b i false ii true iii false
 b and c are in series so will be dimmer than a, which is connected in parallel.

17 a Electricity that builds up and moves across a gap but not along wires.
 b No, because we cannot control the flow of electrons.
 c Electricity that flows along wires.
 d Toaster, refrigerator, power tools, heaters, etc
 e A pathway along which electricity (current) flows.
 f 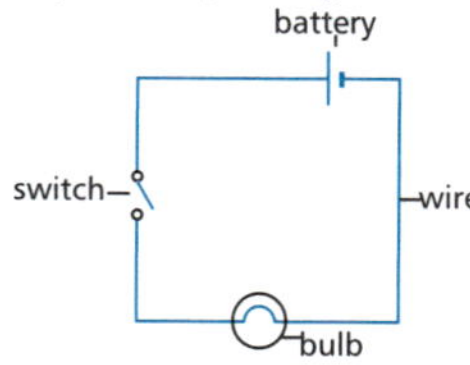

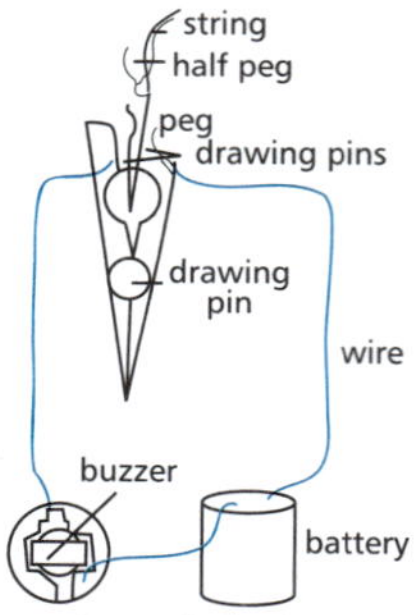

19 fuse, current, circuit

20 It will glow brighter because twice the number of electrons are passing through the filament, giving off twice the amount of energy.

Test B page 95

1 atoms, protons, neutrons, electrons, nucleus, protons, neutrons

2 a element
 b 105
 c hydrogen and oxygen
 d hydrogen, carbon, nitrogen and oxygen

3 a 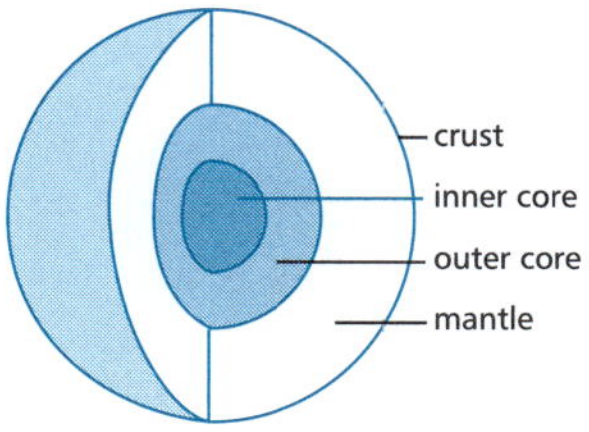

b inner core
c crust
d The further from the core the cooler the layers get.

4 d

5 a close to the surface b granite
c sandstone d basalt

6 C

7 a

Rock	Rock group
marble	metamorphic
sandstone	sedimentary
granite	igneous
slate	metamorphic
basalt	igneous
limestone	sedimentary

b Igneous — formed from cooled and hardened magma from deep in the Earth
Sedimentary — formed when layers of sediment have cemented together
Metamorphic — rock changed by heat, pressure and/or chemical reaction

8 a 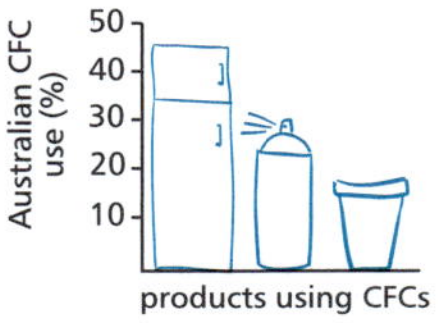

b Because if the layer is thinner more of the sun's energy can pass through.
c stratosphere

9 (1) When rocks come into contact with magma they melt.
(2) Magma solidifies deep under ground or at the crust surface to form igneous rocks.
(3) Surface rocks are broken into bits by the weather.
(4) Bits of rock are carried down mountainsides by rain and creeks.
(5) At the bottom of lakes and seas the bits of rock make sediments.
(6) As the sediments are covered by new layers, sedimentary rock forms.
(7) As sedimentary rocks get buried the temperature and the pressure increase.
(8) Sedimentary rocks slowly change into metamorphic rocks.

10 a light, reflect, light, straight lines, corners, shadow
b stars, fire, light bulb, sun, TV

11 a all b same c y = 45 degrees

12 a red, orange, yellow, green, blue, violet
b A red filter absorbs all the colours in white light except red, so only the red passes through it.

13 a eyes, light, ears, sounds
b Light travels faster than sound.
c Light spreads out more as it gets further away so it is fainter.
d Light can travel through empty space but sound cannot.

14 a Drum skins vibrate which makes the air around them vibrate.
b vibrations, air, ear, brain

6 a refraction
b boundary, normal
c away
d right angles
e Red light is refracted less than the violet light at both boundaries.

16 a tighter b shorter c thinner

17 a yes c close the switch
b no d series

Notes